The nature and causes of climate change

STUDIES IN CLIMATOLOGY SERIES
Edited by Professor S. Gregory, University of Sheffield

A. M. Carleton, *Satellite Remote Sensing in Climatology*

C. M. Goodess, J. P. Palutikof and T. D. Davies, *The Nature and Causes of Climate Change*

B. Yarnal, *Synoptic Climatology in Environmental Analysis*

Other titles are in preparation

The nature and causes of climate change

Assessing the long-term future

C.M. GOODESS, J.P. PALUTIKOF AND T.D. DAVIES

Belhaven Press
London

Lewis Publishers
Boca Raton Ann Arbor

First Published in Great Britain by
Belhaven Press (a division of Pinter publishers),
25 Floral Street, London WC2E 9DS

British Library Cataloguing in Publication Data
A CIP catalogue record for this book is available from
the British Library.
ISBN 1 85293 229 5

First published in the United States of America in 1992 by
Lewis Publishers, 2000 Corporate Blvd., N.W.,
Boca Raton, Florida, 33431

Library of Congress Cataloging-in-Publication Data
Goodess, Clare.
    The nature and causes of climate change / Clare Goodess, Jean
P. Palutikof, and T. Davies.
        p. cm.
      ISBN 0-87371-849-6
      1. Climatic changes.   I. Palutikof, J. P.   II. Davies,
Trevor D.   III. Title.
QC981.8.C5G667   1992   92-14262
551.6--dc20   CIP

Typeset by The Castlefield Press Ltd., Wellingborough, Northants
Printed and bound in Great Britain by Biddles Ltd.

Contents

List of figures

List of tables

List of tables

Acknowledgements

This book arises from a series of studies, undertaken by the authors as part of the UK radioactive waste safety assessment programme, on the implications of future climate change for deep underground disposal. The work has been carried out under contract for UK Nirex Ltd. since 1987 and we would like to thank the organization for making this book possible.

The authors also wish to thank the many individuals who have made this book possible. André Berger and Stan Gregory made valuable comments and suggestions on the reports prepared for UK Nirex Ltd. Stan Gregory recognized the potentially wide appeal of our work and suggested that we contribute to the Belhaven Press Studies in Climatology Series. Special thanks go to Mike Thorne for the thoroughness and enthusiasm with which he read, re-read and improved our manuscripts throughout the long production period. Dennis George and Beki Barthelmie provided welcome support, advice and encouragement over this period. Philip Judge is responsible for the excellent job of redrawing many of the figures. Finally, Clare Goodess acknowledges the support and understanding of Rick Warrens, her family and friends.

The authors have made all reasonable efforts to obtain permission for the use of copyright material. The following individuals and publishers have kindly granted permission for the reproduction of figures and tables:

Professor A. Berger; Figure 2.3 (Figure 4a from Berger, 1978a), Figure 2.6 (Figure 5 from Berger et al., 1990), Figure 2.7 (Figure 26 from Berger, 1988), Figure 5.1 (Figure 9 from Berger et al., 1991b), Figure 9.9 (Figure 8 from Berger et al., 1991b); Professor D.Q. Bowen; Table 7.14 (Table 3 from Bowen and Sykes, 1988), Figure 7.2 (Table 3 from Bowen et al., 1986), Figure 7.3 (Figure 5 from Bowen et al., 1986); Dr R.D. Cess; Figure 4.3 (Figure 1 from Cess and Potter, 1988), Figure 4.4 (Figure 1 from Cess et al., 1989); Dr J.P. Crutchfield; Figure 10.2 (Figure 6a from Tsonis and Elsner, 1989); Elsevier Science Publishers B.V.; Figure 1.3 (Figure 9.1 from Frakes, 1979); Dr C. Genthon and Nature (Copyright (c) 1987 Macmillan Magazines Ltd); Figure 2.8 (Figure 1 from Genthon et al., 1987); Sir John Houghton on behalf of the International Panel on Climate Change, and Cambridge University Press; Table 6.4 (Table 9.1 from Warrick and Oerlemans, 1990), Table 6.5 (Table 9.8 from Warrick and Oerlemans, 1990), Table 6.6 (Table 9.9 from Warrick and Oerlemans, 1990), Table 6.7 (Table 9.10 from Warrick and Oerlemans, 1990), Figure 4.1 (Figure 3 from Houghton et al., 1990), Figure 4.2 (Figure 6 from Houghton et al., 1990); Professor J. Imbrie; Figure 2.5 (Figure 42 from Imbrie and Imbrie, 1979); Professor J. Imbrie and Science (Copyright 1980 by the AAAS); Figure 2.4 (Figure 10 from Imbrie, 1980); Dr P.M. Kelly; Figure 3.5 (Figure 3 from Wigley and Kelly, 1990); Professor J. Kutzbach & Nature (Copyright (c) 1985 Macmillan Magazines Ltd); Figure 3.1 (Figure 1 from Kutzbach and Street-Perrott, 1985); Quaternary Research (Academic Press); Figure 1.1 (Figure 1 from Mitchell, 1976); Dr F. Röthlisberger; Figure 3.2 (from Röthlisberger, 1986); Dr M.J. Tooley and Geographical Journal (Royal Geographical Society); Figure 6.3 (Figure 12 from Tooley, 1974).

Editor's preface

From being a somewhat disregarded minority interest within the field of meteorology, or equally a minority interest in associated fields such as geography, climatology has become — within the last 10 to 15 years — a central focus of innumerable disciplines concerned with the present and future environment of this planet. No longer seen as being concerned solely with average weather, or with general descriptions of recent conditions, it spans time scales from the historical and geological past to probable events in the future; requires massive computing facilities for its statistical and mathematical modelling approaches; draws on the information from and technology of space satellites; and focuses on problems of prime importance to the future well-being of humankind.

Reflecting this growth in scale, quantity, scientific depth and applied relevance of research in climatology has been an equivalent growth in publications. The number of international journals specifically concerned with research papers in this field has increased considerably; conferences on climatic issues have proliferated, organized by a range of official or semi-official bodies, with volumes of proceedings or overall reports tending to be published for each of them; and various series of books concerned with specific aspects of the overall climatological field have appeared.

It is therefore reasonable to ask why another such series should be created, or at least where it fits into the overall picture of contemporary climatology. This series is primarily concerned with global climates at the present, with how they have fluctuated or changed over the past, and with the impact of such conditions on human society. This necessarily includes considerations of data sources, methodologies and theories as well as reviews and discussions of the evidence for the climatic conditions themselves. Such volumes will thus help to delimit and define the framework in the context of which the results of modelling studies, estimates of future climatic conditions, and assessments of environmental change need to be evaluated. So this series does not comprise introductory textbooks, but rather monographs or reviews of the current state of knowledge and understanding, essential for the researcher and perhaps also for the final-year specialist student.

This second volume in the series, written by three authors from the Climatic Research Unit and the School of Environmental Sciences at the University of East Anglia, focuses on the essential theme that has made climatology of such vital present-day concern. The issue of global warming of the atmosphere, resulting from anthropogenically-induced increases in such 'greenhouse gases' as carbon dioxide, methane and many others, has launched climatology into the centre of the international stage, with profound implications for future economic, social and political interrelationships.

This book considers these probable climatic impacts within the framework of longer-term 'natural' changes in climate, both in the past and into the future. The mechanisms that control such changes are reviewed and explained; the modelling approaches that condition scientific thinking about global warming are examined; the possible regional impacts upon the British Isles (taken as a case study) are considered; and the present state of understanding of all these is critically evaluated. Such an interpretative review requires the drawing together of concepts and findings from numerous fields, including meteorology, geology, sedimentology, glaciology, oceanography and biology, a task that has been successfully achieved both in the text and in the comprehensive bibliography.

S. Gregory
19 February 1992

1 Introduction

1.1 The context

The study of climate change is a relatively new and rapidly expanding discipline. Growing awareness of issues such as the enhanced greenhouse effect and the depletion of the stratospheric ozone layer has led to an increased interest in, and funding of, research. The research effort is directed towards understanding the mechanisms of climate change and the prediction of future climate. Substantial and rapid advances in data collection and modelling, and in our knowledge of the climate system, are being made. At the same time, awareness of the uncertainties in prediction is also growing.

Clearly, climate change is not just of academic interest. The potential impacts of climate change on the decadal or century scale have an obvious relevance for future agricultural and economic activity. Information about future climate change is essential for the development of energy policy. Demand for space heating and air conditioning, for example, will depend in part on future temperature conditions. It is highly probable that decisions about appropriate levels of fossil-fuel use will be guided by knowledge of their potential climate impacts.

Environmental impact assessments are an increasingly common feature of the planning process. This reflects the knowledge that the impacts of certain human activities are likely to persist over time scales greater than the 'typical' decadal planning time scale. One example of such an activity is the production and disposal of radioactive waste. The Climatic Research Unit has been involved in a project to assess future climate states in the UK over very long time scales, for input to studies of the performance of underground repositories for low- and intermediate-level nuclear waste (Goodess *et al.*, 1988; 1990; 1992). In the UK, there is no cut-off to the time over which repository safety should be assessed, although our lack of knowledge is such that it is not meaningful to extrapolate millions of years into the future. In the course of the climate project, it was concluded that relatively detailed estimates of future climate states can be made for shorter periods (10 ka[1] to 100 ka, say) and less detailed estimates to 1 Ma, taking into account increasing model and data uncertainties (Goodess *et al.*, 1988; 1990; 1992).

Thus, there is a growing demand for information about future climate over long time scales. In this book we consider climate changes over the next 1 Ma. In order to be able to predict the long-term future, it is necessary to reconstruct the past pattern of climate over equally long time scales, and to identify the forcing mechanisms involved. The extent to which past patterns of change can be used as a guide to the future must be determined. Knowledge of mechanisms of climate change which have operated in the past may also help us to understand future short-term climate change. The climatic effects of the reduced atmospheric CO_2 concentrations observed in

1

previous glacial episodes, for example, may tell us something about the sensitivity of the climate system to the current increase in anthropogenic CO_2.

Generally there is a tendency for climate changes on short and long time scales, and for 'natural' and 'anthropogenic' mechanisms, to be studied as separate issues. A major aim of this book, therefore, is to consider the interactions between these different aspects of climate change.

The growing demand for long-term impact assessments requires the production of information about future climate change at the regional level. While considerable effort is being put into the development of regional scenarios for a high-greenhouse-gas world, rather less attention has been given to the regional patterns of change over longer time scales. A second major aim of this book, therefore, is to investigate how regional patterns of change relate to larger-scale and long-term forcing mechanisms. These relationships are illustrated with the example of one particular region of western Europe, the British Isles.

1.2 Definition of time scales

1.2.1 The time scales of climate variability

Climate varies on all time scales, in response to both random and periodic forcing factors. In Figure 1.1, tentative, notional estimates of the relative variance of the climate from 10 a to 1 Ga, are presented (Mitchell, 1976). This graph is based on interpretation of a combination of palaeoclimatic indicators, such as oxygen isotope records, together with shorter-term historical and instrumental data. It represents a subjective interpretation of the data, not a quantitative assessment.

The spikes or peaks represent the periodic components of variation. If it is assumed that the climate system responds in a linear fashion to combinations of forcing factors, these peaks must correspond to the periodicity of those forcing factors. If, however, the response of the system to forcing factors is strongly non-linear, the periodicities in the response will not necessarily be identical to the periodicities in the forcing factor(s).

The peaks appear above a 'base-line' which represents random variations of climate caused by internal processes and associated feedback mechanisms, often referred to as *stochastic* mechanisms. These mechanisms create what may be thought of as 'white noise' and account for a large proportion of climate variation. An essential corollary of the existence of random processes is that a large proportion of climate variation cannot be predicted. (The implications of chaos theory for the predictability of climate are discussed in Section 10.3.2.)

Figure 1.1 suggests that the strength of the climate response tends to increase with cycle length. The actual climate state at any point in time represents an aggregate response to all cycles of variation superimposed on the background 'noise'.

The dashed line in Figure 1.1 attempts to show the geographic scale of

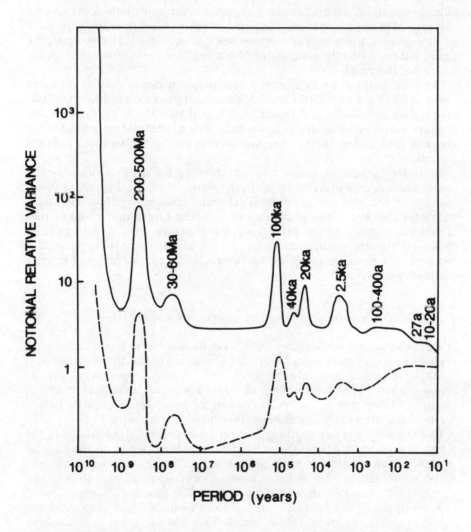

Figure 1.1 Notional estimates of the relative variance of climate over all periods of variation, from 4.5 Ga to 10 a (solid line). The strongest periodic components of variation are labelled. The relative variance for periods in excess of 1 Ga is, in effect, meaningless, since the age of the Earth is insufficiently long for several cycles to have occurred. The dashed line attempts to show the geographical scale of climate variations: the area below it represents the contribution to the total variance of climate processes operating on a spatial scale of less than 1000 km. (From Mitchell, 1976.)

climate variation: the area below it represents the contribution to the total variance of processes operating on a spatial scale of less than 1000 km, such as topographic winds and convective storms. As time scales shorten, the figure indicates that the contribution from regional-scale rather than global variations increases.

On time scales of up to around 1 Ma, major peaks of relative variance occur at 100 ka, 40 ka, 20 ka and 2.5 ka. Lesser peaks are indicated at 100–400 a, 27 a and 10–20 a. At periods of less than 100 ka, the random and regional variations account for upwards of 50% of the total variance. On the shortest time scales, most of the variance is due to random and regional factors.

In order to predict future climate change, we need to identify the mechanisms contributing to the peaks in relative variance. Figure 1.2 shows the major mechanisms of 'natural' climate change together with their characteristic time scales of operation. In order to use the record of past climate as a guide to the future, we must assume that the principle of uniformitarianism applies: that the same set of mechanisms as operated in the past will continue to act in the future, and that the range of climate variability will be similar.

1.2.2 *Identification of an appropriate assessment period*

It would be unwise to use the entire 4.5 Ga history of the Earth's climate as a guide to the climate change over our stated period of interest, the next 1 Ma, even if sufficiently reliable and detailed records could be reconstructed. The contribution of very long-term processes, such as continental drift, can be assumed to be insignificant. An appropriate period for the basis of the assessment must, therefore, be identified.

In Figure 1.3 a generalized temperature and precipitation history of the Earth is presented. These curves have been reconstructed on the basis of evidence from a variety of proxy indicators, including ocean sediments and fossil assemblages. They indicate relative conditions of warmth/cold and dryness/wetness rather than absolute values. The time scale is expanded at the top of the figure because of the greater availability of geological and other evidence for more recent times. This enhanced resolution allows relatively rapid variations to be identified. The last 2 Ma (the Quaternary period) appears unique in the geological record, because of the number of large and rapid variations between, on the one hand, periods of extreme cold, and on the other, 'normal' (close to present-day) temperature values. These are the Quaternary (or Pleistocene) glacial–interglacial cycles. (The Pleistocene includes the entire Quaternary period, with the exception of the most recent 10 ka.) In the glacial periods, present-day temperate regions are invaded by land-based ice caps and sea ice. The glacial–interglacial cycles are accompanied by major fluctuations in sea level and by large-scale changes in plant and animal communities.

If we assume that there will be no major changes in geographical boundary conditions, such as the latitudinal distribution of land areas, then

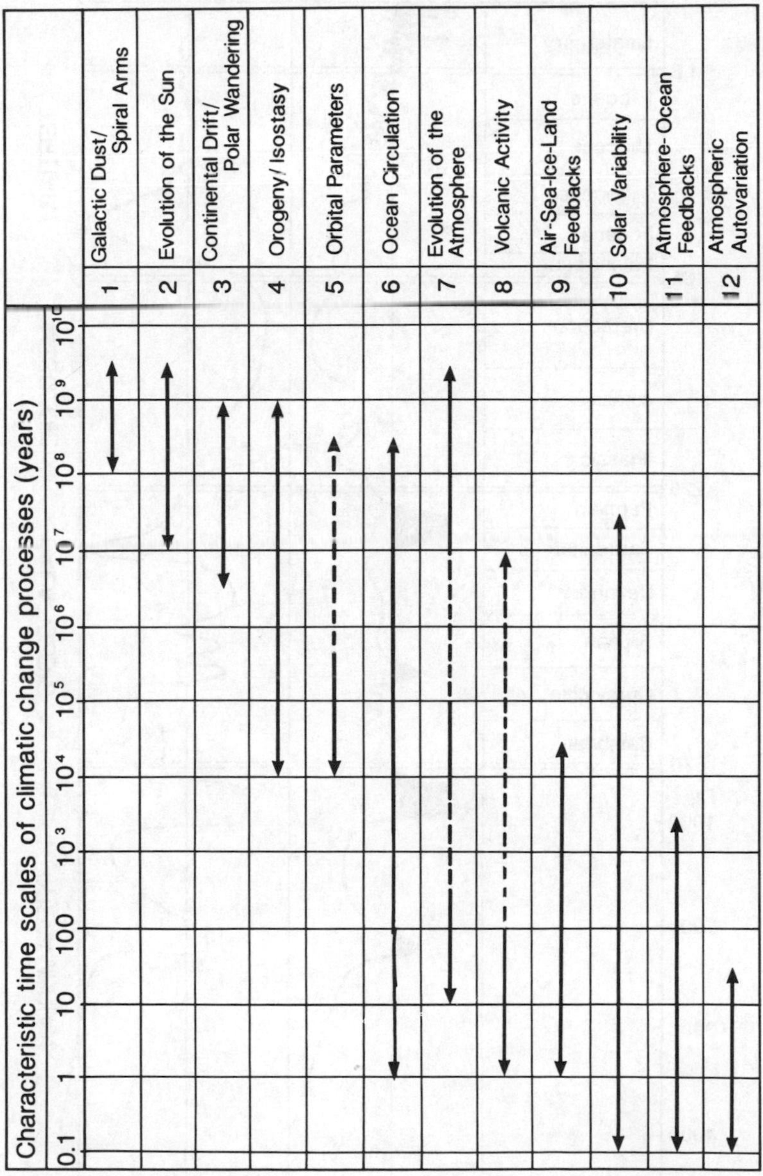

Figure 1.2 Major mechanisms of climate change and their time scales of operation. (Based on Wigley, 1981.)

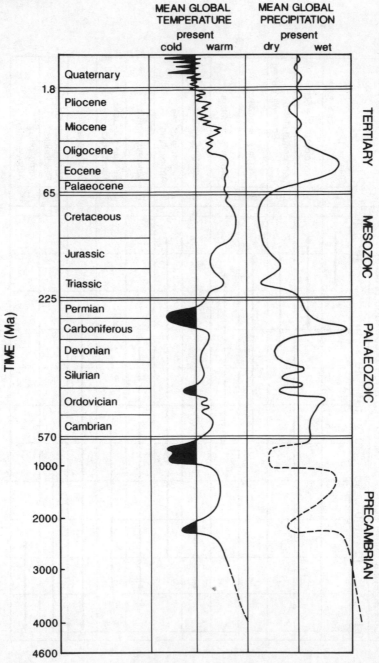

Figure 1.3 Generalized temperature and precipitation history of the Earth. The curves indicate relative departures from present global means. Periods colder than today are shaded. Dashed lines indicate periods of sparse data. Note that the time scale is progressively expanded from bottom to top. (From Frakes, 1979.)

the climatic record of the Quaternary period is likely to provide an appropriate guide to the range of climate change expected over the next 1 Ma. The most probable cause of glacial–interglacial cycles, and of the peaks in relative variance at 20–100 ka identified in Figure 1.1, is considered to be the periodic changes in the Earth's orbit around the Sun (orbital forcing) (Imbrie, 1985; Berger and Tricot, 1986). Based on the assumption of unchanged boundary conditions, the theory of orbital change can also be used to investigate future climate change. Such an analysis must, however, also incorporate the potential effects of anthropogenic climate change.

The assumption about geographical boundary conditions made above appears reasonable, in that continental drift and polar wandering operate on scales of millions of years (Wigley, 1981). The continents have drifted relatively little over the past 20 Ma or so. Tectonic processes such as orogeny (continental uplift and mountain building) operate over equally long time scales. (It is, however, possible that continuing tectonic uplift has produced at least some of the geographical boundary conditions necessary for the initiation of the Pleistocene sequence of glacial–interglacial cycles (see Section 2.4.4).)

For the last 1 Ma, therefore, we can identify a pattern of change (the Pleistocene glacial–interglacial cycles), together with its most likely cause (orbital forcing) and its characteristic time scale (20–100 ka). For the purposes of this book, we define climate change over the orbital time scale as long-term. From this definition, it follows that short-term climate change occurs on scales less than those associated with orbital forcing. We do not consider decadal, or shorter, climate variability in this book. Thus, short-term climate change is defined as change occurring on time scales of 100 a to 20 ka. Short-term changes are superimposed on the longer-term pattern, i.e. on the Pleistocene glacial–interglacial cycles.

The future climate is expected to differ from that of the last 1 Ma in one fundamental respect. It will be influenced by anthropogenic forcing mechanisms. It is likely that the principal climatic influence over the next 1 ka or so will be the enhanced greenhouse effect. A major aim of this book is to consider both the short-term and the long-term effects of anthropogenic change on the 'natural' pattern of climate change.

1.3 Terminology

Because the relationships between regional patterns of change and larger-scale and long-term forcing mechanisms are illustrated in this book with the example of the British Isles, we use the conventional UK terminology for the Pleistocene sequence of climate. In Chapter 7, this sequence is discussed in detail and links between events in the British Isles and in other regions are identified. While we do not wish to pre-empt the discussion in Chapter 7, it is useful to provide some background information for the reader who may not be acquainted with the UK terminology. Table 1.1, therefore, is a provisional guide to British glacial and interglacial periods and equivalent events in northern Europe, the Alps and North America. We stress that this

Table 1.1 Provisional correlation between the British Pleistocene glacial–interglacial sequence and events in northern Europe, the Alps and North America. Interglacial episodes are indicated by italics. *Sources:* Lamb 1977a; Frakes, 1979; Bowen *et al.*, 1986; Fullerton and Richmond, 1986; Richmond and Fullerton, 1986; Sibrava, 1986; Gibbard *et al.*, 1991.

	Britain	Northern Europe	Alps	North America
Late Pleistocene	Loch Lomond Advance	Younger Dryas		Younger
	Late-glacial interstadial	*Alleröd*		
	Devensian	Weichselian	Würm	Wisconsin
	Ipswichian	*Eemian*	*Riss–Würm*	*Sangamon*
	hiatus			
Middle Pleistocene	Wolstonian	Saalian (Warthe/ Drenthe)	Riss	Illinoian
	Hoxnian	*Holsteinian*	*Mindel–Riss*	Yarmouth
	Anglian	Elsterian	Mindel	Kansan
	Cromerian	*Cromerian*	*Günz–Mindel*	Afton
	hiatus			
Early Pleistocene		Bavelian		Pre-Illinoian[2]
	Beestonian			
	hiatus			
	Pastonian			
	Pre-Pastonian			
	Baventian	Menapian[1]	Günz	Nebraskan
	Bramertonian			
	Antian	*Waalian[1]*	*Donau–Günz*	
	Thurnian	Eburonian[1]	Donau	
	Ludhamian	*Tiglian[1]*	*Biber–Donau*	
	Pre-Ludhamian	Praetiglian	Biber	

[1] Based on assessment of evidence from East Anglia and the Netherlands, Gibbard *et al.* (1991) equate all the British events in the Ludhamian to Pastonian sequence with the Tiglian. The Menapian, Waalian and Eburonian events are, therefore, equivalent to the period encompassing the Beestonian and hiatuses in the UK sequence.
[2] It has been recommended that the terms Yarmouth, Kansan, Afton and Nebraskan should be rejected, unless referring to specific sites (Richmond and Fullerton, 1986). An alternative scheme based on sub-divisions of the Pre-Illinoian period is proposed.

table is not a definitive assessment. Particular uncertainty surrounds the Early Pleistocene sequences.

Because of the uncertainties, no attempt is made to assign dates to the events identified in Table 1.1. They are, however, given at appropriate points in the text. With the exception of the Loch Lomond Advance/ Younger Dryas and the Alleröd, stadials (cold periods during interglacials) and interstadials (warm periods during glacials) are also omitted from the table but are discussed in the text.

Table 1.1 covers the Pleistocene period only, i.e. 10 ka BP[2] to approximately 2 Ma BP. The most recent 10 ka period is the Holocene interglacial. We use this generally accepted name in preference to the UK term, the Flandrian.

1.4 The structure

Chapters 2–4 are concerned with the mechanisms of climate change and provide a theoretical background for the discussion of regional climate change in later chapters. First, we consider long-term change. Orbital forcing, as described by the modified Milankovitch theory, is considered to be the major cause of the glacial-interglacial cycles over the Quaternary. In Chapter 2 we review the theoretical, modelling and empirical evidence in support of orbital forcing. This mechanism can only ever explain part of the variability of the palaeoclimate record. Potential mechanisms of short-term climate change, operating over time scales of 100 a to 20 ka, are, therefore, assessed in Chapter 3. As a basis for this assessment, the pattern of climate change over the last 20 ka in the western European and North Atlantic region is described.

Chapters 2 and 3 deal only with 'natural' causes of climate change. Possible changes caused by the enhanced greenhouse effect are discussed in Chapter 4. Particular consideration is given to the reliability of general circulation models (GCMs) which are the most complex and most widely used models in studies of climate change.

Interactions between climate changes induced by natural and anthropogenic forcing mechanisms are expected to occur in both the short and longer term. These interactions are discussed in Chapter 5, with emphasis on possible links between the enhanced greenhouse effect and orbital forcing. Changes in sea level are investigated in Chapter 6.

Chapters 7–9 focus on regional climate change, both past and future. The regional sequences of Quaternary climate are outlined in Chapter 7. A major aim of this book is to understand how regional patterns of climate change relate to larger-scale and long-term forcing mechanisms. To illustrate these relationships, we take as our example one particular region of western Europe, the British Isles. In order to reflect the range of climate variability and data availability within such a region, and to address the needs of climate impact studies in terms of spatial resolution, we consider three areas in detail. These are Caithness in northeast Scotland, Cumbria in northwest England, and Wales. No long continuous reconstructions of past climate are available from these areas. We, therefore, use evidence from elsewhere in northwest Europe to reconstruct the sequence of past climate. In order to extend the analysis of past climate conditions in specific regions beyond the period covered by locally available records, it is necessary to explore correlations between the continuous, fragmentary and often poorly dated regional land-based records and the continuous, mainly marine-based, global and regional palaeoclimate data. The factors inherent in the nature of the different palaeoclimate indicators which complicate such intercomparisons are assessed in Chapter 7.

While Chapter 7 discusses the general pattern of climate change over the Quaternary period in Britain, Chapter 8 provides a more detailed picture of change in the three study areas (Caithness, Cumbria and Wales) over the most recent glacial–interglacial cycle. Two different sources of evidence are used to investigate both the magnitude of change and the prevailing general circulation patterns. The first is GCM output and the second is palaeoclimate data.

Chapter 9 describes the use of GCMs as a guide to future climate conditions in northwestern Europe in a greenhouse gas-induced warming episode. Long-term simulations of future climate produced by orbitally forced climate models are also presented. These modelling results confirm that, anthropogenic forcing apart, the patterns and range of climatic conditions likely to be experienced over the next 1 Ma will be close to those experienced over the last 1 Ma. This implies that it is reasonable to use the reconstructed record of Quaternary climate as a guide to future conditions. These studies also provide an indication of how future climate sequences may diverge from past sequences because of enhanced greenhouse warming.

Finally, Chapter 10 is devoted to a discussion of the major uncertainties associated with long-term assessments of future climate.

Notes

[1] a = annum, ka = thousand years, Ma = Million years, Ga = gigayears, i.e. 10^9 years
[2] BP = Before Present, AP = After Present

2 Long-term climate change: orbital forcing

2.1 Mechanisms

In the mid-nineteenth century, the Scotsman James Croll published an astronomical theory linking the Pleistocene ice ages with periodic changes in the Earth's orbit around the Sun (Croll, 1875). Croll's ideas were refined and elaborated by the mathematician Milutin Milankovitch (1941). The original Milankovitch theory identified the main cause of the Pleistocene ice ages as periodic changes in the distribution of incoming solar radiation, resulting from variations in the Earth's orbital geometry. There are three elements of this geometry which are relevant: *obliquity* or tilt; *precession of the equinoxes;* and *eccentricity*.

Today the Earth is tilted on its rotational axis at an angle of 23.4° relative to a perpendicular to the orbital plane of the Earth. This angle of inclination has fluctuated between about 22° and 24.5° with an average periodicity of 41 ka (Figure 2.1). It determines the latitudinal positions of the polar circles and the tropics. As the angle of tilt, or *obliquity,* increases, so the amount of solar radiation received at high latitudes over the summer season increases. In compensation, winter radiation decreases. Changes in obliquity have little effect at low latitudes, since the strength of the effect decreases towards the Equator. Milankovitch predicted that variations in obliquity would modulate the intensity of the seasonal cycle and alter the strength of the latitudinal temperature gradient. The effect on radiation receipts is equal in each hemisphere.

The axis of rotation of the Earth 'wobbles' because of the gravitational pull of the Sun and Moon on the equatorial bulge of the Earth. Since the Earth's orbit around the Sun is elliptical, this wobble affects the timing of the solstices and equinoxes in relation to the extreme Earth–Sun distances (Figure 2.2). This phenomenon is known as *precession of the equinoxes* and affects the intensity of the seasons. If perihelion (the shortest Earth-Sun distance) occurs in mid-June, i.e. when the Northern Hemisphere is tilted towards the Sun, then Northern Hemisphere summer solar radiation will increase. This situation is illustrated in the bottom part of Figure 2.2. Milankovitch considered that precessional effects have a mean periodicity of 22 ka and that the effects are greatest in low latitudes. The direction of the changes which occur in solar radiation receipt at the Earth's surface is opposite in each hemisphere.

Finally, Milankovitch considered the effects of orbital *eccentricity* caused by the gravitational effects of the Sun, Moon and other planets on the Earth. The Earth's orbit has varied from being near-circular to markedly elliptical with a mean periodicity of about 95.8 ka. These changes modulate the

EARTH-JUNE 21
(NH SUMMER
SOLSTICE)

SUN

EARTH-DECEMBER 21
(NH WINTER
SOLSTICE)

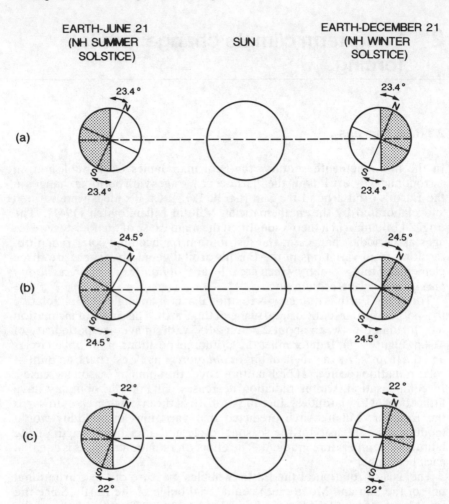

Figure 2.1 The effect of obliquity on the distribution of incoming solar radiation: (a) tilt = 23.4° (present value); (b) tilt = 24.5° (maximum value over a 41 ka cycle); (c) tilt = 22.0° (minimum value over a 41 ka cycle). As tilt increases from (a) to (b), the poles receive more solar radiation. As tilt decreases from (b) to (c), the poles receive less solar radiation. (After Imbrie and Imbrie, 1979.)

effects of precession of the equinoxes. With maximum eccentricity, differences in solar radiation receipt of about 30% may occur between perihelion and aphelion (the smallest and greatest Earth–Sun distances). Since eccentricity modulates the intensity of the seasonal cycle the effects are opposite in each hemisphere.

Orbital geometry variations cause only small variation in the global annual receipt of solar radiation (Berger, 1977): the major effect is on the seasonal and latitudinal distribution of radiation. Milankovitch considered changing seasonal and latitudinal patterns of incoming radiation to be

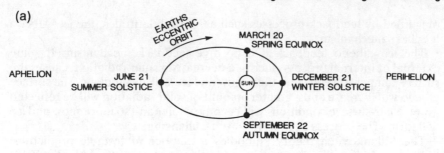

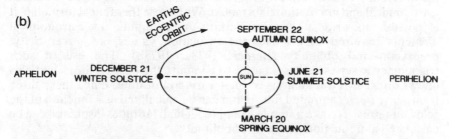

Figure 2.2 Precession of the equinoxes. The equinoxes complete a full cycle approximately every 22 ka. (a) Present-day positions of the Northern Hemisphere equinoxes and solstices. The summer solstice coincides with aphelion (the longest Earth–Sun distance). (b) Positions of the Northern Hemisphere equinoxes and solstices at the opposite phase of the precessional cycle (e.g., 11 ka BP and 11 ka AP). The summer solstice coincides with perihelion (the shortest Earth–Sun distance). (After Imbrie and Imbrie, 1979.)

critical factors in the growth of continental ice sheets and in the initiation of ice ages. Levels of summer radiation at high northern latitudes (60–70°N) were thought to be particularly critical.

Milankovitch identified the orbital configurations thought to be most conducive to glaciation: i.e. minimum obliquity, high eccentricity and aphelion occurring during the Northern Hemisphere summer. He argued that low levels of summer radiation arising from such a configuration would allow the persistence of snow and ice throughout the summer months and into the following winter season, thus allowing ice caps to accumulate. The warmer winters (coinciding with perihelion) would favour increased evaporation from the subtropical oceans, providing more moisture for snowfall. The strong latitudinal temperature gradient resulting in part from low obliquity would, in turn, intensify the atmospheric general circulation and facilitate the poleward movement of moisture. All these factors act to encourage the expansion of ice caps. Milankovitch calculated that such configurations had occurred at 185 ka, 115 ka and 70 ka BP. It should be noted that Milankovitch expected the direct effects of solar radiation to be magnified by feedback processes such as, at high latitudes, the ice–albedo feedback mechanism.

The ice–albedo feedback is a positive feedback mechanism. If some external or internal process, such as a decrease in solar radiation, causes the surface temperature to cool, then snow and ice will accumulate. The surface albedo will increase and a greater amount of solar radiation will be reflected away. Near-surface conditions will become cooler and so more snow and ice will form. The process is, therefore, self-enhancing.

The Milankovitch theory embodies a number of testable predictions linking periodic changes in solar radiation and climate. Milankovitch's original calculations of solar radiation changes based on astronomical and gravitational influences have been refined and expanded (Vernekar, 1972; Berger, 1978a; 1978b; Berger et al., 1989a). Calculations have been made for different latitudes and for different seasons, based on well-established astronomical and gravitational formulae. With these theoretical formulae, it is possible to write a series expansion and compute the astronomical elements required to evaluate the insolation values, i.e. eccentricity, precession and obliquity (Berger, 1978a; 1978b). The use of such trigonometric series makes it easier to compute the desired terms and also allows direct spectral analysis of their variation in time. Once these three terms have been computed, insolation can be calculated as a function of the relevant terms. To take a simple example, at high latitudes insolation can be calculated as a function primarily of obliquity.

In Figure 2.3, some of Berger's computations of variations in Northern Hemisphere insolation during the summer half-year over the last 250 ka are presented. In this season, variations in obliquity (the 41 ka cycle) dominate at high latitudes. Precessional changes (on a 23 ka cycle) dominate at lower latitudes. The relative strength of the three orbital elements varies with latitude and season (Berger and Pestiaux, 1984).

2.2 Geological evidence of orbital/climate cycles

While the reality of solar-radiation changes related to changes in orbital parameters is uncontroversial, there has been considerable controversy concerning the reality of the associated climatic changes predicted by Milankovitch (Imbrie and Imbrie, 1979). The Milankovitch theory has been tested by comparing the geological evidence of climate change with the predicted cycles of change. It is only relatively recently that sufficiently long geological records have become available, and could be dated with the required accuracy. Pleistocene chronologies were not firmly dated until the early 1970s, when new theories and evidence of reversals of the Earth's magnetic field became available (Hays and Berggren, 1971; Shackleton and Opdyke, 1973).

2.2.1 Evidence from deep-sea cores

The most significant evidence in support of the Milankovitch theory has

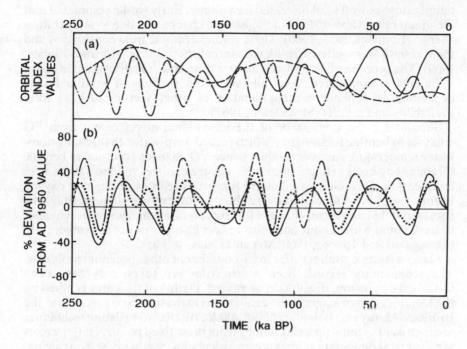

Figure 2.3 Orbital changes and solar insolation curves. (a) Variations of eccentricity (- - -), obliquity (——), and precession (– - – -), over the last 250 ka. (b) Northern Hemisphere summer solar radiation at 80°N (—), 65°N (·····), and 10°N (– - – -) expressed as percentage departures from 1950 values. (After Berger, 1978a.)

come from studies of the chemical composition of deep-sea cores. Oxygen isotope studies have been particularly important. During colder periods, the ratio of ^{18}O to ^{16}O in water vapour evaporated from the oceans is relatively small. In turn, condensation in the atmosphere under colder conditions produces precipitation with a relatively small ratio of ^{18}O to ^{16}O. Thus, in colder periods precipitation is relatively enriched in ^{16}O, and a fraction of this ^{16}O-enriched precipitation is trapped in the cryosphere. During glacial periods, the oceans are, therefore, relatively enriched in ^{18}O, with high ^{18}O to ^{16}O ratios in seawater indicating a relatively high ice volume.

The shells of calcareous marine fauna fall to the sea floor when they die and the oxygen isotope content of these calcium carbonate deposits reflects the oxygen isotope content of the ocean at the time of deposition. Analysis of shells in deep-sea sediments should, therefore, allow past isotope ratios to be calculated. The process of marine faunal deposition itself (calcium carbonate crystallization), however, results in a slight excess concentration of ^{18}O in the calcium carbonate relative to the water. The strength of this enrichment process depends on temperature: the excess ^{18}O increases as the temperature decreases. Thus it might appear that greater concentrations of ^{18}O in calcium carbonate deposits can be used as a measure of lower ocean temperature as well as of increased ice volume. Early studies estimated that the greater part (about 70%) of the isotope change was due to temperature change (Emiliani, 1955; 1966). These studies aroused great controversy and their results were often considered unrealistic (Dansgaard and Tauber, 1969). The proposed link between temperature and ^{18}O enrichment in ocean sediments is considered doubtful, since the process of marine faunal deposition is dependent on a number of other factors such as food availability and salinity (Shackleton, 1967).

Support for the dominance of the ice volume effect comes from ^{18}O analyses of benthic foraminifera which record deep-water changes. Bottom-water temperatures are very stable, hence ^{18}O changes recorded in benthic foraminifera must reflect, primarily, changes in ice volume rather than temperature. Estimates based on the temperature interpretation of oxygen isotope ratios have tended to overestimate temperature changes. It is now considered that about two-thirds of the isotope-ratio changes reflect changes in ice volume while about one-third reflect changes in ocean temperature (Dansgaard and Tauber, 1969; Mix and Pisias, 1988).

Oxygen isotope studies suffer from a number of other potential problems. The sedimentary records from which cores are taken may have been disturbed by erosion, dissolution or mixing. Dating of the cores is based on the identification of magnetic reversal events in the longest cores, such as the Brunhes–Matuyama boundary about 700 ka BP. Relative dating techniques are then used to build up a chronology from these fixed points. Dating errors are likely if sedimentation rates are miscalculated. Studies based on calcium carbonate analysis may be misleading if the influence of locally occurring turbidity currents, which tend to dilute the ^{18}O-rich calcium carbonate laid down in glacial times, is ignored.

Despite these interpretational problems, the value of marine sediments as palaeoclimatic indicators is widely recognized for three main reasons. First, they provide long and continuous records. Second, records derived from the sediments tend to produce a fairly consistent picture of globally synchronous climate fluctuations. Cores from different parts of the world can be cross-correlated with relative ease. Finally, it is possible to date core samples reasonably accurately, through a combination of absolute and relative dating techniques. For these reasons, marine cores have provided the stratigraphic framework for the Quaternary era and produce compelling evidence in support of the orbital theory.

Spectral analysis of isotopic records from ocean cores has identified climatic cycles coinciding approximately with the Milankovitch precession (22 ka) and obliquity (41 ka) cycles. As longer and more accurate records have become available, they have resulted in some slight modifications to Milankovitch's original predictions.

The 782 ka record of ice volume provided by core V28-238, from the Pacific Ocean (Shackleton and Opdyke, 1973), shows a dominant 100 ka cycle together with lower amplitude cycles at about 40 ka and 20 ka (Figure 2.4). The 100 ka isotope cycle (Figure 2.4b) is approximately in phase with the 95.8 ka eccentricity cycle (Figure 2.4a). The dominance of the eccentricity cycle was not predicted by Milankovitch, who thought that it

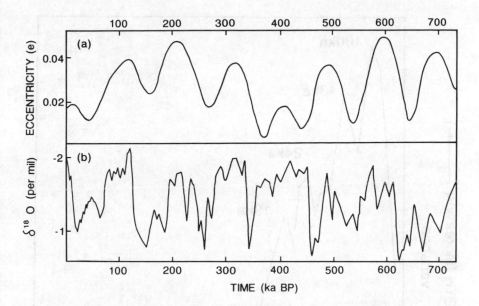

Figure 2.4 Eccentricity and global ice volume record for the past 730 ka. (a) Variations in orbital eccentricity (Berger, 1977; 1978a). (b) Oxygen isotope curve from deep-sea core V28-238 (Shackleton and Opdyke, 1973). (From Imbrie and Imbrie, 1980.)

would only be significant in its interaction with precessional effects. The 100 ka cycle is also seen in an isotopic record from two Indian Ocean cores (Hays *et al.*, 1976). Spectral analysis of these two cores reveals an obliquity cycle of 43 ka, not 41 ka (Figure 2.5). Given the accuracy of geological dating methods, this difference is not significant. The precession cycle appears as two peaks: a major one at 24 ka; and a minor one at 19 ka. The independent identification of this double peak in the geological data (Hays *et al.*, 1976) and in the orbital and insolation data (Berger, 1977) provided the earliest most compelling evidence in support of the orbital theory. Similar cycles have also been identified in the 2.1 million year record from core V23-239 from the western Equatorial Pacific (Shackleton and Opdyke, 1976).

Thus, by the early 1980s, spectral analysis studies had established a statistical link between orbital changes and the geological record of climate change (Shackleton and Opdyke, 1973; 1976; Hays *et al.*, 1976; Imbrie and Imbrie, 1979). The major periodicities and effects of the orbital elements, as understood at this time, are summarized in Table 2.1. Recent years have seen major data collection and modelling efforts by the international research community, aimed at advancing understanding of the relationships between orbital forcing and climate.

While the underlying astronomical theory remains unchanged, some refinements have been made to the theoretical calculations of past orbital

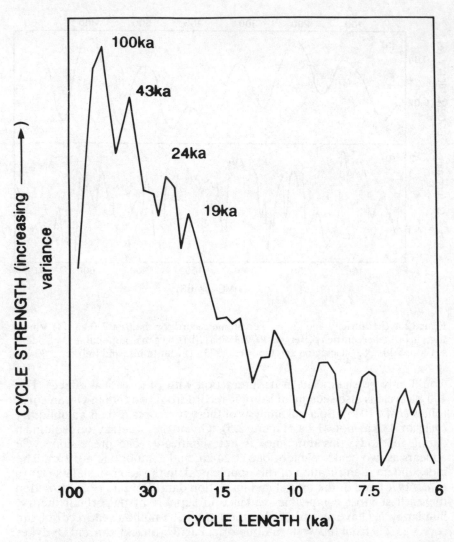

Figure 2.5 Spectral analysis of a 500 ka record from two Indian Ocean cores. Peaks of high relative variance are labelled. (Data from Hays *et al.*, 1976; figure from Imbrie and Imbrie, 1979.)

and insolation changes. These calculations have been revised a number of times (Berger, 1978a; 1984; 1989a; Berger *et al.*, 1989a; 1989b). Calculations are now available for the last 10 Ma), although these are less certain before 7.5 Ma BP.

In addition to the major precessional periodicities of 23 ka and 19 ka, secondary spectral peaks occur at around 17 ka, 15 ka and 56 ka (Berger, 1989a). The main obliquity period of 41 ka is accompanied by minor cycles

Table 2.1 The major orbital elements and their effects on climate.

Orbital element	Present index value	Index range	Average periodicity (ka)
Obliquity (ϵ)			
Strength of effect increases towards the poles if caloric seasons are considered. Effect is equal in both hemispheres.			
decreasing *increasing*	$\epsilon = 23.4°$	$\epsilon = 22°$ to 24.5°	41
lower seasonality, steeper insolation gradient increased summer high-latitude radiation, intensified seasonality, reduced insolation gradient			
Precession (ω)			
Variations alter structure of seasonal cycle by changing Earth–Sun distance in each season. Effect is complex and modulated by eccentricity, parametrized as index $\Delta e \sin\omega$. High positive $\Delta e \sin\omega$ is associated with lower Northern Hemisphere summer insolation.	$\omega = 0.01635$	$\omega = 0.05$ to -0.05	19 23
Eccentricity (e)			
Causes 0.2% variation in annual incoming radiation. Main effect is to modify amplitude, but not frequency, of precession cycle (modifies duration and intensity of seasons). Over winter and summer seasons, effects are greatest at low latitudes and are opposite in each hemisphere.	$e = 0.0167$	$e = 0.0005$ to 0.0607	95 (main spectral components at 410, 95, 120 and 100)

of length 53 ka and 30 ka. The major eccentricity periodicities are 413 ka, 95 ka and 136 ka. Secondary peaks are located at 13 ka, 2.3 Ma, 1.3 Ma and 600 ka. Finally, periodicities of 59 ka and 64 ka years have been identified for the combined effects of precession and obliquity. It is argued that the identification of a number of these secondary orbital periodicities in the geological record (Ruddiman *et al.*, 1986; Stothers, 1987) provides further support for the Milankovitch theory (Berger, 1989a).

Advances in drilling, dating and analysis techniques, both on land and in the oceans, continue to be made. Major international research projects such as the Ocean Drilling Program are increasing the spatial distribution of reliable, long-term geological records. New analysis techniques are also being developed. While the earliest ocean-core studies, for example, concentrated on fossil assemblages as indicators of water mass types, and on

^{18}O as an indicator of global ice volume, more recent studies have looked at a wider range of ecological and chemical indicators. These include Cd/Ca ratios and the ^{13}C content of foraminifera, which are indicative of changes in the nutrient availability, productivity and vertical structure of the oceans (Boyle and Keigwin, 1987; Boyle, 1988; Sarnthein et al., 1988; Duplessy et al., 1989; Mix, 1989). Many of these new studies have been important in improving our knowledge of climate processes, feedbacks and cause-and-effect mechanisms, particularly those in operation over the last glacial–interglacial cycle. Changes in deep-water production and the thermohaline circulation are, for example, now thought to play an important role in the processes of glaciation and deglaciation, together with changes in the atmospheric concentrations of greenhouse gases such as carbon dioxide and methane (see Section 2.4).

2.2.2 Other palaeoclimate evidence

Evidence of climate cycles at the major Milankovitch periodicities between 19 ka and 413 ka is now available from ocean sediments (Hays et al., 1976; Shackleton et al., 1988; Imbrie et al., 1989; Shackleton and Imbrie, 1990), land-based sediments (Herbert and Fischer, 1986; Olsen, 1986; Winograd et al., 1988; Beget and Hawkins, 1989; Kukla and An, 1989), pollen cores (Broecker et al., 1988), ice cores (Genthon et al., 1987) and coral reef sea-level records (Mesolella et al., 1969; Chappell and Shackleton, 1986). Here we review some of the more recent land- and marine-based evidence.

The presence of climate cycles in Chinese loess sediments has been long recognized. Recent advances in the technique of magnetic susceptibility dating have allowed direct links to be made with marine oxygen isotope chronologies (Hovan et al., 1989; Kukla and An, 1989). The magnetic susceptibility of loess and the interbedded soils varies with the concentration and grain size of magnetic minerals and with pedogenesis (Kukla et al., 1988).

Loess layers form during periods of relatively cold and semi-arid climate conditions which are characterized by a poorly vegetated semi-desert environment (Kukla et al., 1988; Kukla and An, 1989). Such conditions over central China are typical of glacial periods. Magnetic susceptibility measurements provide the most accurate and objective method of distinguishing individual soil and loess layers. Susceptibility is highest in the soil horizons and lowest in the loess horizons. Susceptibility records could be interpreted in terms of fluctuating temperature or wind velocity but, in the case of the Chinese sediments, are more closely related to variations in effective soil moisture and hence regional precipitation (Kukla and An, 1989).

The thickness ratios of soil to loess layers in the Chinese sediments are comparable to the thickness ratio of warm (odd-numbered) and cold (even-numbered) stages identified in ocean sediment records (Kukla et al., 1988). It is, therefore, reasonable to assume that the magnetic minerals in loess are deposited at a constant rate. On this assumption, it is possible to interpolate between the magnetic reversal horizons, which are reliably dated global events, to develop a time series (Kukla *et al.*, 1988).

Generally good agreement has been found between the Chinese loess sequences of the last 500 ka and the downwind oxygen isotope record from core V21-146 in the North Pacific Ocean (Hovan *et al.*, 1989). This core is tuned to the SPECMAP time scale developed by Imbrie et al. (1984). Variations in the dust content of core V23-146 reflect climate variations over the Chinese source area: more dust is produced, transported by wind, and deposited in the North Pacific Ocean during cold, dry glacial periods than during warmer periods. Similar orbital periodicities can be identified in both the land-based and marine-based records. The 100 ka eccentricity cycle, for example, dominates the loess record for the last 500 ka, whereas the shorter orbital periodicities dominate the period 2.4–0.5 Ma BP (Kukla and An, 1989). Strong evidence for a 400 ka cycle has also been detected in Chinese loess records for this period (Wang *et al.*, 1990).

Magnetic susceptibility variations in loess have also been used to identify, for the first time, orbital periodicities in land-based wind-blown sediments at high latitudes (Beget and Hawkins, 1989). The loess layers are interpreted as being associated with periods of higher wind intensities which are typical of colder, glacial climate conditions. Significant correlations between a 250 ka loess record from central Alaska and a high-resolution marine oxygen isotope record, tuned to the SPECMAP time scale (Martinson *et al.*, 1987), have been found. The highest correlations are found when the loess record is assumed to lag behind the marine record by about 10–11 ka. Spectral analysis of the loess sequence reveals peaks at 125 ka, 41 ka and 23 ka (Beget and Hawkins, 1989). The strongest peak is that associated with the 41 ka obliquity cycle. The dominance of this peak at higher latitudes was predicted by Berger (1978a).

Orbital periodicities have also been identified in proxy records of the intensity of the African and Asian monsoons (Prell and Kutzbach, 1987; Bloemendal and de Menocal, 1989). These proxy records include tropical lake levels, the terrigenous wind-blown content of ocean sediments, Mediterranean sapropols (layers of organic-rich sediments), the pollen content of sediments in the Gulf of Aden and Indian Ocean sea-surface temperatures. Prell and Kutzbach (1987) employed a number of such records in a study of monsoon variability over the last 150 ka. Four periods of strong African and Asian monsoons were identified at around 10 ka, 80 ka, 105 ka and 125 ka BP, closely coinciding with periods of maximum Northern Hemisphere summer radiation (Berger, 1978a). The observational and modelling evidence presented by Prell and Kutzbach suggests that, over the last 150 ka, the monsoon mechanism may have been particularly sensitive to the precessional cycle. However, the response of the monsoon mechanism to orbital forcing during glacial periods is less distinct. In the Last Glacial Maximum, for example, precipitation decreased over southern Asia but increased over equatorial Africa. It is concluded that both orbitally induced radiation changes and glacial boundary conditions influence the major regional features of the monsoons (Prell and Kutzbach, 1987).

Analysis of core 721 in the western Arabian Sea, drilled as part of the Ocean Drilling Program, revealed periodic changes in the strength of the Asian monsoon over the last 3.5 Ma (Bloemendal and de Menocal, 1989). The magnetic susceptibility record of core 721 is considered to be a good indicator of variations in the transport of terrigenous material by the Asian monsoon wind systems. Increased wind transport indicates a stronger monsoon, and vice versa. Variability is found to be concentrated at the Milankovitch frequencies (Bloemendal and de Menocal, 1989). The 400 ka and 100 ka eccentricity cycles are seen to modulate the response of the monsoon system to the shorter obliquity and precessional cycles. During the last 2.4 Ma, the 41 ka obliquity cycle dominated. Prior to this time, before the onset of Northern Hemisphere glaciations, the precessional cycles (19 ka and 23 ka) were stronger.

Cycles with periodicities close to those associated with the Milankovitch theory have now been identified in a number of pre-Quaternary sedimentary records as far back as the mid-Palaeozoic about 400 Ma BP (Fischer, 1986; Berger, 1989b). The possibility that Milankovitch forcing has occurred over 'at least the last 2.2 Gyr of Earth history' has recently been suggested (Berger, 1989b). There are inevitable methodological problems associated with the identification of periodicities for the remote geological periods. Precise dating is difficult and, in some cases, it may not be possible to identify a mechanism linking the orbital/insolation changes with sedimentary changes.

These pre-Quaternary cycles are of particular interest because they do not involve growth and decay of the major Northern Hemisphere ice sheets, which first appeared about 2.5 Ma BP (Loubere and Moss, 1986; Shackleton et al., 1988; Dowsett and Poore, 1990). Evidence suggests that the mid-Cretaceous (100 Ma BP), for example, was a relatively warm period (global temperatures up to 12° higher than today), with higher atmospheric CO_2 concentrations and no large-scale continental glaciations. Spectral and time-series analysis of a 100 Ma record of carbonate content from a sequence of black shales in the Italian Appennines indicates a concentration of variance at the Milankovitch periodicities (Herbert and Fischer, 1986). The 100 ka and 400 ka eccentricity cycles dominate the record and evidence is also found of the higher-frequency precession cycle. Similar results for the Cretaceous have been found by a number of other researchers in the Appennine sediments, in chalk sediments of the American western interior and in marine sediments in the Atlantic (see Berger, 1989b, for a review of the evidence). Cycles with a mean periodicity of 23.5 ± 4.4 ka have been detected over at least 20 Ma in seven marine cores from the South Atlantic (Herbert and D'Hondt, 1990). These apparent cycles in Cretaceous and other pre-Quarternary non-glacial periods raise a number of questions about the mechanisms linking orbital forcing and the climate response. These are discussed in Section 2.4.

Evidence is also emerging which reveals temporal variability in both the length and the strength of the orbital periodicities. For example, recent calculations of the orbital forcing over very long time scales show that, because of slow changes in the Earth's rotation rate and ellipticity, together with changes in the Earth–Moon distance, the orbital periodicities are not constant (Berger *et al.*, 1989a; 1989b). Geological data and astronomical models are available which allow investigation of the effect of changes in the Earth–Moon distance and day length (a measure of the changing rotation rate of the Earth) over the last 440 Ma BP on the obliquity and precessional cycles (Table 2.2). The length of the obliquity and precessional cycles is found to shorten back in time because of the decreasing Earth–Moon distance and day length. The greatest influence is on the obliquity cycles. Reliable geological data and models of the Earth's rotational velocity and related parameters are not available prior to 440 Ma BP, but approximate theoretical estimates of these parameters can be made over longer periods. On this basis, precession and obliquity periodicities can be estimated for the last 2.5 Ga (Berger *et al.*, 1989a). The tendency towards longer obliquity and precessional periodicities as the present day approaches is seen over the entire period.

Table 2.2 Estimates of the changing length of the orbital periodicities back in time.

Geological epoch	Ma BP	Precession		Obliquity	
		19 ka	23 ka	41 ka	54 ka
Holocene	0	19 000	23 000	41 000	54 000
Late Cretaceous	72	18 645	22 481	39 381	51 226
Early Permian	270	17 638	21 034	35 145	44 284
Late Carboniferous	298	17 421	20 725	34 291	42 936
Middle Devonian	380	16 824	19 886	32 053	39 484
Early Silurian	440	16 399	19 296	30 546	37 222

Source: Berger *et al.*, 1989a.

These theoretical calculations have not, so far, been substantiated by the pre-Quaternary geological evidence. No reduction in the length of the precession or obliquity cycles back in time has yet been observed and the periodicities of the eccentricity cycles seem to be unchanged over time. This may, however, be a reflection of dating uncertainties and other interpretational problems.

While there is no direct observational evidence for temporal changes in the length of the various orbital periodicities, there is substantial evidence for changes in their strength. Over the most recent 500 ka, for example, the 100 ka cycle dominates the Chinese loess records whereas in the earlier part of the record the higher orbital frequencies are more important (Kukla and An, 1989). In the marine oxygen isotope record, the obliquity cycle dominates during the period 2.5–1.1 Ma BP (Shackleton et al., 1988).

Other Quaternary records reveal a gradual apparent increase over time in the variance of the climate system associated with orbital forcing. In the sea-surface temperature and oxygen isotope records from the North Atlantic cores K708-7 and DSDP 552A (Ruddiman *et al.*, 1986; Shackleton *et al.*, 1988), the amplitude of the 100 ka cycle has increased by a factor of 4 over the last 1 Ma. This increase was fastest during the period 700–400 ka BP and has been accompanied by an increase in the amplitude of the precession cycles. The concurrent increases in the amplitudes of the eccentricity and precession cycles imply that the strength of the response to eccentricity is modulated by the strength of the precession cycle. The increased amplitude of the orbital cycles is reflected primarily in a tendency towards more severe glacial episodes (colder sea-surface temperatures and greater ice volume) (Shackleton *et al.*, 1988).

The available observational evidence thus indicates temporal variations in the strength of the orbital periodicities, which are apparently reflected in the response of various components of the climate system. There is also evidence of spatial variability in the response to orbital forcing during the Quaternary. A comprehensive study of this variability is being carried out by members of the SPECMAP group (Imbrie *et al.*, 1989). This is the group which developed a stacked chronology tuned to the record of orbital changes (the SPECMAP time scale), using five marine oxygen isotope records (Imbrie *et al.*, 1984). The SPECMAP time scale is widely used as a standard Quaternary stratigraphy.

The SPECMAP group's latest study involves analysis of the spatial pattern, amplitude and phase of sea-surface temperature (SST) response to each of the three orbital parameters, and comparison with oxygen isotope ice volume records (Imbrie *et al.*, 1989). The SST data are derived from faunal assemblages recorded at 17 core sites from 54°N in the Atlantic to 44°S in the Indian Ocean. The SPECMAP time scale (Imbrie *et al.*, 1984) is used for all cores. Evidence of the orbital periodicities is found at all sites, but there are marked differences in the amplitude and phase of SST response (Imbrie *et al.*, 1989). For all three orbital parameters, the SST response in high southern latitudes leads that of high northern latitudes. The same pattern is seen in the ice volume records although, in this case, the lead in the Southern Hemisphere response to the obliquity cycle is not considered to be significant (Imbrie *et al.*, 1989).

Spatial variability is also seen in the relationship between the SST and the ice volume responses to orbital forcing (Imbrie *et al.*, 1989). In the Southern Hemisphere, the SST changes lead the ice volume response to precession, whereas in the Northern Hemisphere the SST changes lag the ice volume response. The greatest lags occur in the region of the subtropical gyre. The lags and leads in the response to precession are all considered to be significant, as are those associated with the SST response to obliquity (Imbrie *et al.*, 1989). In the latter case, however, they are smaller. Clear differences are also seen in the response of the two hemispheres to eccentricity. The Southern Hemisphere response is latitudinally dependent. Between 1°N and 30°S the observed leads and lags are not consistent. Beyond 30°S the SST response leads the ice volume response. In the Northern Hemisphere, the SST response is in phase with the ice volume response.

The analysis of spatial variability in the frequency, amplitude and timing of the SST response is part of an ongoing research effort to understand the chain of causal mechanisms linking the response of the climate system to orbital forcing (Imbrie *et al.*, 1989). We discuss the implications of the identified leads and lags in Section 2.4. First, we consider the ability of climate models to reproduce the major characteristics of the Quaternary record identified in this chapter.

2.3 Modelling the Quaternary glacial–interglacial cycles

In this section we consider the range of models which have been developed to aid investigation of the reality of orbital forcing. In particular, we are interested in what these models can tell us about the nature of the climate system response to orbital forcing. Is it linear or non-linear? Is all the forcing external? (Orbital forcing is considered to be external because it affects the distribution of radiation entering the climate system at the top of the atmosphere.) What is the role, if any, of internal forcing? Does orbital forcing drive the climate system or act as a pace-maker of change?

Three groups of models can be identified: time-dependent ice sheet models; statistical/numerical dynamic models; and energy balance and general circulation models. Tables 2.3 and 2.4 summarize the major characteristics of a number of models from the first two categories. These have been reviewed in detail elsewhere (Berger et al., 1984; Imbrie, 1985; Berger, 1988). Some have been used to simulate future climate (Kukla and Kukla, 1972; Calder, 1974; Weertman, 1976; Imbrie and Imbrie, 1980; Berger et al., 1981; Kukla et al., 1981; Oerlemans and Van der Veen, 1984). Their projections are discussed in Section 9.3.

The ice-sheet models have a number of common characteristics (Broecker and Denton, 1989). A two-dimensional profile through a Northern Hemisphere land-based ice sheet is modelled. Orbital forcing is represented by summer insolation receipts at high latitude, typically about 65°N. Ice flow is represented by an approximate flow law which ignores factors such as basal sliding. The northern margin is fixed while the southern margin and thickness vary with the accumulation and ablation rates, which are themselves sensitive to elevation. The resulting ice volume changes are compared with marine oxygen isotope records in order to assess model performance.

Table 2.3 Characteristics of time-dependent ice-sheet models forced by orbital changes.

Reference	Model characteristics
Weertman, 1976[1] Pollard et al., 1980 Watts and Hayder, 1984	Linear response models. 100 ka cycle is poorly reproduced.
Oerlemans, 1980 Birchfield et al., 1981	Models include a time lag for isostatic deformation.
Pollard, 1982; 1983a; 1983b; 1984	Model includes an ice-calving term to accentuate process of ice-sheet retreat.
Peltier and Hyde, 1984[2] Hyde and Peltier, 1985; 1987[2] Peltier, 1987[2]	Model emphasizes role of glacial isostasy.
Peltier, 1988a	Model includes sea-level change and a standard ice-sheet model of Antarctica.

[1] Details of this model are given in Table 9.2.
[2] Models described in the text.

In early model experiments (Weertman, 1976; Pollard *et al.*, 1980), the response to forcing is linear and due mainly to precession and obliquity. The 100 ka cycle is very weak and the melting of ice during glacial terminations is underestimated. In later models, a variable time lag for isostatic loading and unloading is introduced but the observed strength of the 100 ka cycle is still underestimated and some models appear over-sensitive to changes in the value of model parameters (Oerlemans, 1980; Birchfield *et al.*, 1981). The addition of mechanisms which enhance melting, such as ice-sheet calving, can produce more realistic 100 ka cycles (Pollard, 1982; 1984; Watts and Hayder, 1984).

A more recent model incorporates an improved scheme for glacial isostasy which takes into account mantle viscosity (Peltier and Hyde, 1984; Peltier, 1987). The latest version is able to reproduce a reasonable 100 ka cycle over the last 500 ka, but is less successful in reproducing the record of the earlier period between 800 ka and 500 ka BP. The 100 ka model cycle has a sawtooth shape reflecting the typical slow growth and rapid melting of ice sheets (Peltier, 1987; Hyde and Peltier, 1987). The realistic representation of the rapid termination of glacial episodes is attributed to a two-stage mechanism, which reflects the greater sensitivity of ice sheets to changes in ablation as opposed to accumulation. First, during the glacial episode itself, the stability of the ice sheet decreases as it increases in area and height. As the model climate warms at the end of the glacial period, the ablation zone becomes progressively and disproportionately larger. In the second stage, a critical level of warming is reached and a large ice-free area develops on land equatorward of the former ice-sheet margin, in a region which is still depressed due to glacial loading. Ice from the poleward area of accumulation then begins to slide and surge into this depression and rapidly melts.

The 100 ka cycle of ice-sheet build-up and decay simulated by the model is not caused directly by the eccentricity forcing (Hyde and Peltier, 1987). Since the strongest eccentricity forcing occurs at a point in each climate cycle when the model ice sheets respond purely to precession, and because ice sheet decay is triggered once orbitally induced warming reaches a critical level, it is considered that eccentricity phase-locks the climate response to precession. Because of the presence of additional orbital periodicities (obliquity and the lower-frequency eccentricity cycles) the phase-locking is not perfect (Hyde and Peltier, 1987). This model is, therefore, an example of an internally-driven system, the oscillations of which are phase-locked by external forcing.

In contrast, results from a model incorporating ice sheets, but with different representations of the relevant physical processes, suggest that the external orbital forcing, together with internal forcing, drives the observed 100 ka cycles (Le Treut and Ghil, 1983). This model consists of a non-linear climate oscillator which includes radiation balance, ocean thermal inertia, a very simplified hydrological cycle, the mass balance and plastic flow of ice sheets, elasticity of the Earth's crust and mantle viscosity. In the absence of orbital forcing, the model produces self-sustained oscillations of length 5–15 ka. These are caused by interactions between the positive ice–albedo feedback (see page 14) and the negative precipitation–temperature feedback (as temperature rises so does evaporation leading to increased precipitation and, hence, to increased ice-sheet accumulation). The interaction of these two feedbacks within the model is modified by a third feedback mechanism which incorporates the response to bedrock loading and ice–mass balance. When orbital forcing is introduced, the orbital frequencies are amplified and, because of the non-linearity of model interactions, combination tones between the various forcing frequencies appear. Power is transferred from the precession band and a strong 100 ka cycle is produced. This is an example of a non-linear system driven by external and internal forcing.

We see that the most recent ice-sheet models are capable of reproducing the broad pattern of sawtooth glacial–interglacial cycles found in the geological record of the last few hundred thousand years. They indicate that the climate response to orbital forcing is non-linear and that both external and internal forcing is involved. Despite the improvements in the performance of ice-sheet models, a number of problems remain (Denton *et al.*, 1986; Broecker and Denton, 1989). First, the models are unrealistic. Land-based ice sheets alone are considered, although marine ice sheets did expand during previous glacial episodes. It has been suggested that marine ice-sheet dynamics and sea-level changes play an important part in linking and controlling ice-sheet development and decay (Denton *et al.*, 1986). Second, the models show obvious errors. In some model reconstructions of the last glacial termination, for example, the Greenland ice sheet is destroyed (Birchfield *et al.*, 1981), although in fact it is still present today (Denton *et al.*, 1986). Third, these ice-sheet models leave out many of the proposed mechanisms linking orbital forcing and the climate response. These mechanisms include changes in atmospheric composition and ocean circulation changes (see Section 2.4).

The second major group of models is that of statistical/numerical dynamic models. The basic characteristics of a number of these are given in Table 2.4.

The first five models listed in this table have been used to predict future climate (see Section 9.3). Their principal characteristics, together with details of model calibration and reliability, are given in Table 9.2. Output from each model indicates at least general agreement with the geological record. However, spectral analysis of model results and marine oxygen isotope records reveals shortcomings. Two of the models, for example, exaggerate the amount of power at the precessional periodicities of 23 ka and 19 ka compared with that at 41 ka, the obliquity periodicity (Calder, 1974; Imbrie and Imbrie, 1980). Neither of these two models, nor the Kukla *et al.* (1981) model, produce sufficient power at the 100 ka eccentricity period. Other shortcomings have been recognized, such as excessive parameter sensitivity (Table 9.2). The particular models described here have been calibrated with the geological record back to, at maximum, 782 ka BP, and it is noticeable that correlations are reduced beyond about 350 ka BP in some cases (Imbrie and Imbrie, 1980; Kukla *et al.*, 1981).

None of the five reflects the observed changes in the sensitivity of the climate system to the various types of orbital forcing, nor the strength of the

Table 2.4 Characteristics of statistical/numerical dynamic models forced by orbital changes.

Reference	Model characteristics
Kukla and Kukla, 1972[1]	Rates of insolation change are calculated and 'positive/negative insolation regimes' identified.
Calder, 1974[1]	Ice volume grows in proportion to deficit of radiation below a certain point and melts in a different proportion above that point.
Imbrie and Imbrie, 1980[1]	Simple non-linear model simulates lag between orbital variation and ice-sheet response using four adjustable parameters to tune model to geological record 0–150 ka BP.
Berger *et al.*, 1981[1]	Autoregressive multivariate spectral model. Climate is a function of insolation and climate of last 3 ka.
Kukla *et al.*, 1981[1]	Combines three orbital parameters in time-lag bivariant model. The ACLIN and ACLIN 1 formulae describe the link between orbital perturbations, insolation and climate.
Pisias and Shackleton, 1984[2]	Based on Imbrie and Imbrie, 1980. Includes changes in atmospheric CO_2.
Saltzman and Sutera, 1984[2] Saltzman *et al.*, 1984[2] Saltzman and Maasch, 1988[2] Maasch and Saltzman, 1990[2]	Various versions of this model are described in the text. Basically, it consists of linked dynamical equations representing the departure of three components of the climate system from equilibrium. The model is run with and without orbital forcing in order to establish the role of internal and external forcing.
Snieder, 1985	Based on Imbrie and Imbrie, 1980.
Neeman *et al.*, 1988a; 1988b	Zonally averaged statistical/dynamical hemispheric climate model. Includes topography and detailed sea ice/snow and land snow schemes.

[1] Full details of model given in Table 9.2.
[2] Model described in text.

observed 100 ka cycle (Imbrie, 1985; Mix, 1987). They treat the response to orbital forcing as derived from an externally driven non-linear system. Pisias and Shackleton (1984) introduce an element of internal forcing to the model of Imbrie and Imbrie (1980) in the form of the changes in atmospheric CO_2 concentration recorded over the last glacial–interglacial cycle (see Section 2.4.2). This modification improves the simulation of both the spectral characteristics and the phase lags between orbital forcing and ice volume changes.

The different versions of the dynamic model developed by Saltzman and colleagues suggest that the 100 ka cycle is attributable to a free oscillation of an internally-driven non-linear system in which external orbital forcing acts as a pace-maker (Saltzman and Sutera, 1984; Saltzman *et al.*, 1984; Saltzman and Maasch, 1988; Maasch and Saltzman, 1990). The earlier version of the model consists of linked dynamic equations which represent the departure of three components of the climate system from equilibrium (Saltzman *et al.*, 1984). A terrestrial ice-sheet component is represented by the mass of the deep continental ice sheets. The second component is the mass of all marine ice sheets, and the third is a non-cryospheric climate component represented by the mean temperature of thc world's oceans. Processes such as sea-level change and bedrock depression are not, therefore, explicitly included in this model. In the total absence of orbital forcing, the model produces an approximate 100 ka cycle, although it is much more regular and cyclical than that observed in the marine oxygen isotope record. The phase of the model variability is apparently arbitrary and no high-frequency variability is present. In the second stage of the model experiment, forcing from the three orbital parameters is introduced to the dynamic system. The model is run from an arbitrary initial state at 1 Ma BP. By about 500 ka BP, the model response becomes coherent and appears to lock into a stable pattern. In comparison with the model run without orbital forcing, the shape of the 100 ka cycle over the last 500 ka BP is more realistic, and higher-frequency variability at the precessional and obliquity periodicities appears Agreement with the SPECMAP geological record is reasonable. The authors conclude that orbital forcing acts as a pace-maker to the free oscillation of an internally driven non-linear system (Saltzman *et al.*, 1984). This model differs from that of Le Treut and Ghil (1983) in that 100 ka oscillations occur in the absence of any orbital forcing.

In the most recent experiments, three new variables are combined in a similarly constructed dynamic model (Saltzman and Maasch, 1988; Maasch and Saltzman, 1990). The three variables represent global ice mass, atmospheric CO_2 concentration, and deep-ocean warmth and salinity (Saltzman and Maasch, 1988). The experimental methodology is the same as in the previous studies, but orbital forcing is represented by July insolation at 65°N rather than by the linear combination of all three orbital parameters. Without orbital forcing, and using a time constant of 10 ka for ice-sheet response, 100 ka cycles of ice mass and CO_2 are produced as part of a natural oscillation involving deep ocean temperature and salinity. When orbital forcing is introduced, reasonably good agreement between ice mass and the SPECMAP oxygen isotope record and between CO_2 and the Vostok record of past CO_2 changes (see Section 2.4.2.) is achieved (Saltzman and Maasch, 1988). It is concluded that, without orbital forcing, the CO_2 cycle provides the instability needed to drive a natural oscillation involving feedbacks between the cryosphere, atmosphere and ocean (Maasch and Saltzman, 1990).

In the latest experiment with this model, changes in the character of the climate oscillations recorded in the Pleistocene geological record are explored (Maasch and Saltzman, 1990). In particular, a possible cause of the emergence of the 100 ka cycle is sought (see Section 2.2). It is shown that, in the absence of orbital forcing, slow linear changes in the model control parameters, representing tectonic changes, can lead to abrupt changes in the climate response when a critical value is reached. When orbital forcing is added to the model the gradual emergence of the 100 ka cycle over the Late Pleistocene is reproduced. As in previous experiments, the higher-frequency oscillations become apparent and the cycles are phase-locked so that glacial terminations occur at the correct time. Plausible causes of the control parameter changes are suggested (Maasch and Saltzman, 1990). These include progressive growth or erosion of the ocean floor in the North Atlantic and Pacific which could modify the effect of ice on North Atlantic Deep Water production (see Section 2.4.3), and slow uplift of the Tibetan Plateau and western North America which could influence the atmospheric circulation and induce cooling in some regions (see Section 2.4.4.).

This later model (Maasch and Saltzman, 1990) is particularly interesting because it is able not only to reproduce the amplitude and phase of the 100 ka cycle, but also to simulate its emergence. Although this model is based on parameters which are considered to be physically plausible, it is highly simplified. None of the models from the two groups described above can be considered as a fully realistic representation of the whole climate system. Nor is any capable of explaining precisely how the orbital forcing produces dominant 100 ka cycles in marine- and land-based geological records from so many different parts of the world. What these models do confirm is that the response to orbital forcing is non-linear and that it involves some element of internal forcing. Whether the external orbital forcing drives the internal forcing (Le Treut and Ghil, 1983), phase-locks the oscillations of an internally driven system (Hyde and Peltier, 1987), or acts as a pace-maker to free oscillations of an internally driven system (Saltzman and Maasch, 1988; Maasch and Saltzman, 1990) is, however, still uncertain (Imbrie *et al.*, 1989).

The third group of models comprises energy balance models (EBMs) and general circulation models (GCMs). EBMs are relatively simple one-dimensional models of the Earth's radiation balance. GCMs are essentially three-dimensional models of the climate system developed from meteorological forecasting models. Their major characteristics are discussed in Section 4.4. Palaeoclimate GCM experiments fall into two categories: snapshot and sensitivity studies (Street-Perrott, 1991).

Snapshot experiments are designed to investigate the nature of climate conditions and circulation patterns at selected times in the past for which substantial quantities of palaeoclimate data are available and/or which have distinctive insolation forcing characteristics. Experiments are, therefore, mainly restricted to episodes during the last 18 ka, notably the Last Glacial Maximum (18 ka BP) and the Holocene thermal optimum (6 ka BP). In snapshot experiments many of the characteristics of the palaeoclimate data are reproduced, but these model studies are limited in what they can reveal about the mechanisms linking orbital forcing and climate. This is because model boundary conditions, such as ice-sheet location and size, and sea-surface temperature, are prescribed from the palaeoclimate data. These studies do, however, provide useful information for the investigation of temperature, precipitation and atmospheric circulation changes associated with particular orbital configurations and boundary conditions (see Section 8.2). Sensitivity studies have also been performed to assess the relative influence of changed boundary conditions on the climate system and are discussed in Section 8.2.

The majority of palaeoclimate studies carried out with GCMs and EBMs are equilibrium studies. They do not model the time-dependent behaviour of the climate system, which is crucial if we are to understand the processes of glacial initiation and decay. Even where EBMs are used to model time-dependent behaviour, many important components such as snow and ice feedbacks, clouds and the deep oceans, may be grossly simplified or ignored. Thus they can only be used to investigate isolated elements of the climate response to orbital forcing, such as the differential response of land and sea (Short *et al.*, 1991). Our discussion of orbital forcing and associated climatic modelling indicates the need for a physically realistic model of the time-dependent behaviour of the coupled climate system, including the atmosphere, oceans (surface and deep water), cryosphere (marine- and land-based), lithosphere and biosphere. Unfortunately, such a model is likely to be too complex, given the constraints of computing power and speed. A sectorially averaged time-dependent physical climate model is, however, being developed by Professor Berger and colleagues at the Université Catholique de Louvain in Louvain-la-Neuve, Belgium. The current version of this model takes into account feedbacks between the atmosphere, the upper ocean, sea ice, ice sheets and the lithosphere (Berger *et al.*, 1990).

The model consists of a two-dimensional climate model (Gallée *et al.*, 1991) coupled with an ice-sheet model. The seasonal cycle of the Northern Hemisphere is simulated by the climate model which includes a mixed-layer model of the ocean. The climate model is asynchronously coupled to dynamic models of the Greenland, northern American and Eurasian ice sheets, and is run in time steps of 1 ka. Model-generated estimates of variables such as temperature, sea ice, snow depth and ice-sheet elevation are used as inputs to each successive time step. Thus the model generates its own boundary conditions. In the palaeoclimate experiments, the only forcing applied is that of insolation receipts for each day of the year for each latitude band of the Northern Hemisphere.

The model has been used to simulate the volume of the three major Northern Hemisphere ice sheets over the last interglacial–glacial cycle (Figure 2.6; Tricot *et al.*, 1989; Berger *et al.*, 1990). Agreement with oxygen isotope (Figure 2.7; Labeyrie *et al.*, 1987) and sea-level records (Berger, 1988; Berger *et al.*, 1991b) is generally good. The current version of the model, however, overestimates the volume of the Greenland and North American ice sheets during the most recent part of the Holocene (about 6 ka BP to date). In contrast, the North American ice-sheet volume at the Last Glacial Maximum is underestimated (Tushingham and Peltier, 1991). The simulated climate is found to be very sensitive to the initial size of the Greenland ice sheet, the positive ice–albedo feedback, the negative precipitation–elevation feedback over ice sheets, the parametrization of snow albedo and the insolation increase. The model has also been used to

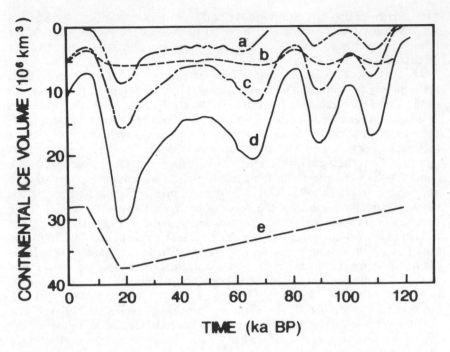

Figure 2.6 Variations over the last interglacial–glacial cycle of ice volume simulated by the Berger *et al* (1990) climate model for (a) the Eurasian ice sheet; (b) the Greenland ice sheet; (c) the northern American ice sheet; and (d) the total of the three Northern Hemisphere ice sheets. The dashed line (e) shows the estimated ice volume of the Antarctic ice sheets interpolated from volumes at 18 ka BP and today. (From Berger *et al.*, 1990.)

simulate future climate; results from these experiments are discussed in Chapters 5 and 9.

A number of improvements to the Louvain-la-Neuve model are under way, such as coupling the climate and ice-sheet models with deep-ocean and global carbon-cycle models (Berger *et al.*, 1991a). Although the results presented here may be considered only preliminary, they confirm the results of many of the simpler and physically less realistic models discussed above. Orbital forcing is confirmed as a highly plausible cause of the Quaternary glacial–interglacial cycles. The modelling evidence also implies that internal processes and feedbacks, such as ice-sheet dynamics and changes in atmospheric composition, play an important part in the climate response.

The importance of ice-sheet dynamics and other internal processes and feedbacks is confirmed by orbital change experiments performed with a linear EBM asynchronously coupled with a vertically integrated ice-sheet model (Deblonde and Peltier, 1991a; 1991b). In the first set of experiments, continental ice-sheet growth over the last glacial–interglacial cycle is simulated (Deblonde and Peltier, 1991a). The location and volume of

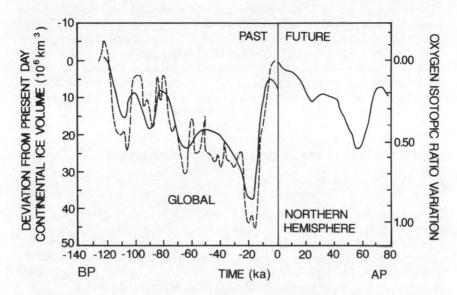

Figure 2.7 Variations in total continental ice volume, over the Earth from 140 ka BP to present, and over the Northern Hemisphere from present to 80 ka AP, as simulated by the Berger *et al.* model (solid line). The dashed line shows the variation in oxygen isotope ratios given by Labeyrie *et al.* (1987) and Duplessy *et al.* (1988). (From Berger, 1988.)

individual Northern Hemisphere ice-sheet masses at the Last Glacial Maximum are reproduced satisfactorily. Surface temperatures are, however, too low over Scandinavia and the Barents and Kara Seas. These discrepancies are related to the lack of ocean currents in the model. More seriously, the coupled model fails to predict the termination and collapse of the ice sheets. This failure is related to the non-establishment of warm North Atlantic ocean currents in the interglacial period and also to the neglect of other factors such as changes in CO_2, CH_4, dust loading, sea level and Southern Hemisphere climate (Deblonde and Peltier, 1991a).

In the second set of experiments, potential causes of the emergence of the 100 ka cycle in the geological record of the last 900 ka, and its dominance over the last 400 ka, are explored (Deblonde and Peltier, 1991b). These two features are reproduced by adding an additional feedback mechanism which is dependent on the time rate-of-change of ice volume. This feedback is defined as a 'generalized meltwater parametrization' (GMP). It could be interpreted as the sudden switching on of a marine-induced instability in continental ice volume, possibly in association with the reorganization of the ocean general circulation. The GMP simulates catastrophic ice-sheet melting, which is triggered when a critical value of the rate of ice volume decrease is exceeded. Once this occurs, the ice-sheet equilibrium line (above which accumulation dominates, and below which melting dominates) is raised sharply and rapid melting occurs.

By introducing a GMP to the coupled model at 800 ka BP and gradually increasing its strength to 400 ka BP, reasonably good agreement with the SPECMAP record and Ocean Core ODP 677 is achieved (Deblonde and Peltier, 1991b). The physical causes of the GMP mechanisms and ice-sheet instability need to be identified. Potential candidates include pro-glacial lakes, ice calving, tectonic uplift, changes in the thermohaline circulation and changes in the atmospheric circulation. These cause-and-effect mechanisms are considered in the next section.

2.4 Internal forcing and cause-and-effect mechanisms

2.4.1 Defining the problem

Early attempts to associate causal mechanisms with glaciation were relatively simple. One of the widely accepted hypotheses of glaciation (Ruddiman and McIntyre, 1981a) states that ice-sheet growth is favoured when Northern Hemisphere summer insolation levels are low as a result of orbital changes, and high-latitude oceans are warm in winter, thus providing an abundant moisture supply. In these circumstances, ice is not melted in summer and builds up in winter until positive feedback mechanisms such as the ice–albedo feedback come into play to enhance ice growth. A variant of this hypothesis describes the role of feedbacks in deglaciation (Ruddiman and McIntyre, 1981b; Jones and Ruddiman, 1982). An increase in Northern Hemisphere summer radiation begins to melt the ice sheets. As a result, there is an increased flow of icebergs and meltwater to the oceans. Icebergs and meltwater depress sea-surface temperature, thus decreasing evaporation and reducing moisture supply for snowfall. Meltwater has a lower salinity than seawater and freezes relatively easily in winter, so further suppressing moisture supply and starving the ice sheets. Both the glacial meltwater and melting icebergs would cause an increase in sea level, thus leading to increased calving and break-up on the continental ice-sheet margins.

As the geographical distribution, dating and diversity of past climate and ocean records have improved, evidence has emerged which fails to fit the theories discussed above (Denton *et al.*, 1986; Broecker and Denton, 1989; Imbrie *et al.*, 1989). These discrepancies can be listed under six headings. First, orbital changes have produced synchronous changes in mountain glaciation and snowline elevation throughout the world (Broecker and Denton, 1989). Second, the 100 ka climate cycle is seen in a wide range of land- and marine-based records (see Section 2.2). Third, events such as the onset of the last glacial termination about 14 ka BP seem to have been more or less simultaneous, although summer insolation forcing had directly opposing effects in northern and southern mid-latitudes (about 45°N and 45°S). Fourth, orbital forcing produces latitudinal variations in insolation, but regional climates do not always appear to reflect local forcing. Fifth, it has been suggested that the development or decay of Northern Hemisphere ice sheets has an impact on global climate (Denton *et al.*, 1986). However, modelling studies indicate that the climatic influence of even the largest recorded North American and Eurasian ice sheets is restricted to the Northern Hemisphere and that the Antarctic ice sheet does not have a significant effect on the Southern Hemisphere (Manabe and Broccoli, 1985). Furthermore, evidence from low- and mid-latitudes of the Pacific and Atlantic oceans shows that the Southern Hemisphere leads the response to orbital forcing (Imbrie et al., 1989). Finally, there are questions concerning the emergence of the 100 ka cycle as the dominant cycle over the last 500 ka and the existence of orbital periodicities in the pre-Quaternary geological record.

Any proposed explanation of the mechanisms linking orbital forcing and climate response must be able to account for the geological evidence. The search for further information has been aided by a variety of new and improved techniques such as ice coring, the analysis of ancient air bubbles trapped within polar ice, and the use of new tracers such as Cd/Ca ratios. In the rest of this section, we consider the potential mechanisms linking cause-and-effect, involving the atmosphere, the oceans and geographic boundary conditions.

2.4.2 The atmosphere

The first time series of changes in atmospheric CO_2 content, based on analysis of ancient air bubbles trapped in ice cores from Antarctica and Greenland, were published in the early 1980s (Berner et al., 1980; Delmas et al., 1980; Neftel et al., 1982). The present-day concentration of atmospheric CO_2 is 353 parts per million by volume (ppmv) (Houghton et al., 1990). The ice-core measurements show that CO_2 levels during the Last Glacial Maximum (18 ka BP) fell to about 180–240 ppmv (approximately 50–70% of the present level) and rose rapidly during the period 16–10 ka BP to a level of about 280 ppmv (about 80% of the present level).

These results have been confirmed by more recent results from Antarctica, notably those from the 160 ka Vostok core (Barnola et al., 1987; Genthon et al., 1987; Jouzel et al., 1989; Lorius et al., 1990) and the 50 ka Byrd core (Neftel et al., 1988). A temperature record is also available for the Vostok core, based on analysis of the deuterium content (Jouzel et al., 1987). The good agreement between the CO_2, temperature, insolation and oxygen isotope marine records suggests that all are related in some way (Figure 2.8). As we discuss below, study of the leads and lags between the various records supports the role of CO_2 variability as a positive feedback mechanism (Genthon et al., 1987; Lorius et al., 1990).

Analysis of the Vostok ice core shows that, during warming episodes, CO_2 and temperature changes occur almost simultaneously, but lag the insolation changes. The same relationships are seen in the Dye 3 core from Greenland (Stauffer et al., 1984). In cooling episodes, CO_2 and temperature changes again lag the insolation changes, but the CO_2 changes significantly lag the temperature changes. These phase relationships confirm that CO_2 may provide a positive feedback rather than a driving mechanism and suggest that different processes cause the decrease and the increase in CO_2 during glaciation and deglaciation. Identification of these processes is,

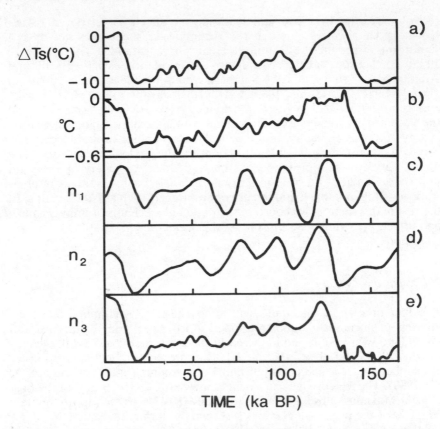

Figure 2.8 Time series based on analysis of the Vostok ice core: (a) temperature change; (b) atmospheric CO_2 expressed as radiative forcing; (c) 65°N July insolation; (d) ice volume from model of Imbrie and Imbrie (1980); and (e) ice volume change from deep ocean cores. Curves c–e are scaled to have approximately the same amplitude. (From Genthon *et al*. 1987.)

however, still controversial. Many of the mechanisms which have been proposed involve changes in marine biological productivity, which can now be estimated from sedimentary and foraminiferal species data (Mix, 1989). As marine productivity increases, the concentration of metabolically produced CO_2 in the oceans increases and the atmospheric concentration of CO_2 decreases.

The theories put forward initially involved changes in nutrient availability. It was suggested, for example, that the erosion of nutrient-rich organic sediments on continental shelves exposed as sea level fell could increase the nutrient content of the ocean surface and so lead to increased productivity during glacial episodes (Broecker, 1981; 1982). Estimates of nutrient availability can now be made from measurements of Cd/Ca ratios in marine sediments (Boyle and Keigwin, 1987), but there is no evidence of large-scale changes in the nutrient content of surface waters during the last glacial episode (Keigwin and Boyle, 1989).

Other theories have invoked changes in the vertical structure and productivity of the ocean as plausible mechanisms (Mix, 1989). Siegenthaler and Wenk (1984), for example, proposed that ocean circulation changes, such as an increase in low-latitude upwelling, could affect the distribution of chemical properties and hence the CO_2 content and uptake of the surface ocean. Evidence from the last interglacial–glacial cycle provides growing support for changes in the vertical structure of the oceans (Boyle, 1988; Duplessy *et al.*, 1989; Mix, 1989; Duplessy and Juillet-Leclerc, 1991).

Analysis of Cd/Ca ratios and the $\delta^{13}C$ content of benthic foraminifera suggests that the concentration of nutrients and metabolic CO_2 and, hence, the region of greatest marine productivity, shift from intermediate to deep-water layers of the Atlantic Ocean, and possibly also the Pacific Ocean, during glacial times (Boyle, 1988) As this redistribution starts, Doyle argues, the concentration of dissolved CO_2 in deep water increases. Initially, because of the excess CO_2 in the deep ocean, the dissolution of carbonate from shells increases. Eventually, the rate of dissolution exceeds that required to maintain the steady-state carbonate input–output balance. The alkalinity of the ocean increases until steady-state dissolution rates are restored. According to Boyle's argument, the atmospheric concentration of CO_2 decreases in direct response to the increase in ocean alkalinity.

The various models which have been proposed imply that large but geographically varying changes in ocean productivity should be observed. Productivity can be estimated from organic carbon accumulation rates in deep-sea sediments (Sarnthein *et al.*, 1988) or from the planktonic foraminiferal content of deep-sea sediments (Mix, 1989). Results from both types of analysis indicate that higher productivity occurred in many places during the Last Glacial · Maximum. There are, however, some interpretational problems associated with the first type of analysis, including possible contamination by organic matter from continental sources. Here we discuss a study based on analysis of planktonic foraminifera in Atlantic sediments (Mix, 1989). Transfer functions linking observed present-day productivity with the faunal assemblages in the core tops were developed. These were calibrated against a model of global productivity and then applied to the entire core sequence.

The results of this study show that productivity was about 40% higher during the Last Glacial Maximum than today over the whole Atlantic, and up to 90% greater than today near the Equator. If these changes are extrapolated to all the oceans then it appears that a significant proportion of the observed CO_2 changes may be driven by changes in productivity (Mix, 1989). An explanation of the productivity changes is still, however, needed. Mix suggests changes in equatorial upwelling and vertical mixing in the subtropics as two possible mechanisms.

Although uncertainty still surrounds the mechanisms by which CO_2 concentrations have varied over the last interglacial–glacial cycle, there is little doubt that such changes did occur. Modelling evidence supports the role of CO_2 as a positive feedback mechanism (Pisias and Shackleton, 1984; Genthon *et al.*, 1987; Adem, 1988; Saltzman and Maasch, 1988; Harvey, 1989a; Lindstrom and MacAyeal, 1989; Rind *et al.*, 1989). Using an energy balance model, Harvey (1989a), for example, finds that the decrease in atmospheric CO_2 at 18 ka BP enhanced global cooling by up to 1.5°C. Genthon *et al.* (1987) estimate that the direct radiative effect of the CO_2 changes over the last glacial–interglacial cycle, as recorded in the Vostok core, is 0.6°C. The direct effect is amplified by various feedback mechanisms such as the water vapour feedback. In this positive feedback mechanism, cooling leads to reduced evaporation and lower concentrations of atmospheric water vapour which, because water vapour is itself a greenhouse gas, in turn leads to further cooling. The operation of this feedback during a period of anthropogenic greenhouse warming is discussed in Section 4.4.3. Results from a multivariate analysis of the Vostok CO_2 and temperature records, using a range of amplification factors, imply that the CO_2 changes could account for 27–85% of the observed temperature variance (Genthon *et al.*, 1987).

Analyses of air bubbles in the Vostok and Byrd cores, both from Antarctica, and the Dye 3 core from Greenland, reveal changes in the atmospheric concentration of methane (CH_4), which also acts as a greenhouse gas. The evidence shows that methane concentrations almost double, from about 350 parts per billion by volume (ppbv) to 650 ppbv, during the transition from glacial to interglacial conditions (Raynaud *et al.*, 1988; Stauffer et al., 1988). The Vostok CH_4 record is correlated with the CO_2 and temperature records from the same core. Spectral analysis reveals similar orbital periodicities in all three records (Chapellaz *et al.*, 1990).

Although the CH_4 and CO_2 records from the Vostok core are correlated, the processes controlling the variations of the two gases are not the same. Whereas the variations in CO_2 are considered to be largely controlled by changes in ocean circulation and chemistry, the CH_4 changes are likely to reflect terrestrial processes. It is thought that the CH_4 fluctuations principally reflect fluctuations in the area of wetlands (Chapellaz *et al.*, 1990). Changes in the extent of low-latitude wetlands associated with changes in monsoon intensity may be particularly important. It is also possible that changes in temperature have a direct effect on CH_4 fluxes from wetlands.

As a greenhouse gas, CH_4 has a direct radiative effect, but also has indirect photochemical effects involving ozone and water vapour. Allowing for both direct and indirect effects, Chapellaz *et al.* (1990) estimate that CO_2 and CH_4 changes may have accounted for about 2.3°C of a total global warming during the last glacial–interglacial transition of 4.5±1.0°C. Another study, based on multivariate analysis of the Vostok records, suggests that the total contribution of greenhouse gas fluctuations to the regional warming seen in the Vostok temperature record (Figure 2.8) is 3°C, or 50 ± 10% (Lorius *et al.*, 1990). From these two studies, it follows that CO_2 and CH_4 may be responsible for 50% of the observed glacial-to-interglacial warming. Methane alone may account for 10% of the warming. Although the available data are not yet reliable, the possibility exists that changes in the atmospheric concentrations of other greenhouse gases such as N_2O may also have occurred during the last glacial-to-interglacial transition (Zardini *et al.*, 1989).

The Vostok records provide further information about changes in the aerosol content of past atmospheres (Legrand *et al.*, 1988a; 1988b; Petit *et al.*, 1990). The atmospheric loading of aerosols from marine and terrestrial sources is generally greater during colder periods, such as 100–15 ka BP, than during warmer periods, such as 130 ka and 10 ka BP (Legrand *et al.*, 1988a; 1988b). The increased loading is related to increases in both the sources and transport of aerosols during glacial episodes. Higher wind speeds, stronger meridional circulation, the growth of arid continental areas and the exposure of continental shelves due to sea-level fall all contribute towards the observed aerosol increase (Legrand *et al.*, 1988a; 1988b; Petit *et al.*, 1990). Falling concentrations of terrestrial dust recorded in the Vostok core lead Antarctic temperature increases during the last glacial–interglacial transition (Petit *et al.*, 1990) so that, by the time of mid-transition, dust loadings are back to low levels.

The increased aerosol loading during glacial periods has a positive feedback effect. The particles reflect radiation back to space, thus enhancing surface cooling. Using an energy balance model, Harvey (1989a) estimates that the increased loading associated with the Last Glacial Maximum may have caused an additional cooling of 2.2°C.

Legrand *et al.* (1988a) have examined the potential causes and effects of changes in the concentration of marine aerosols. Dimethylsulphide (DMS) is produced by planktonic algae and oxidizes in the atmosphere to form a sulphate aerosol called marine non-sea-salt (*nss*) sulphate (Charlson *et al.*, 1987). The concentration of marine *nss* sulphate in the Antarctic atmosphere can be estimated from the Vostok core sulphate record (Legrand *et al.*, 1988a).

Initial analyses of the data suggested that the concentration of marine *nss* sulphate in the Antarctic atmosphere increased by 20–46% during the Last Glacial Maximum (Legrand *et al.*, 1988a). A complete, and more detailed, record of biogenic sulphur emissions over the last glacial–interglacial cycle is now available from the Vostok ice core (Legrand *et al.*, 1991). This record indicates that the concentration of marine *nss* sulphate, corrected for changes in the accumulation rate, varied from 100–106 ng g^{-1} in interglacial periods (the Ipswichian and Holocene), to 136–162 ng g^{-1} in cold periods (prior to 141 ka BP; 65–56 ka BP; and 33–13 ka BP). Methanesulphonate, which is primarily derived from oceanic sulphate, also varied, between about 5 ng g^{-1} in the warm periods and 16–21 ng g^{-1} in the cold periods.

Legrand *et al.* (1988a) believe that the peak concentrations of *nss* sulphate in the Vostok core are due not only to more efficient poleward atmospheric transport, but also to increased productivity. The record shows a strong spectral peak at 21 ka and a weaker 42 ka peak. The existence of these peaks at the orbital frequencies, and the correspondence with the Vostok CO_2 record, suggests a link between CO_2, ocean productivity and DMS. Legrand *et al.* (1988a) propose that this link could be a change in the surface ocean circulation which leads to increased marine productivity and, hence, to increased DMS production and a reduction in atmospheric CO_2 concentrations.

It has been suggested (Charlson *et al.*, 1987) that increased DMS production leads to increased atmospheric concentrations of cloud condensation nuclei (CCN). The increased CCN concentrations in turn lead to an increase in cloud albedo, thus more radiation is reflected back to space. Since increased marine productivity leads to surface cooling, the proposed feedback effect would be positive during glacial episodes. DMS production and its potential role in a period of greenhouse gas-induced warming is discussed further in Section 4.4.

Assuming that the proposed feedback mechanism is plausible, and that concentrations of CCN increase in direct proportion to marine *nss* sulphate, Legrand *et al.* (1988a) have estimated the likely upper limit of the effect during the Last Glacial Maximum. It is estimated that a 46% increase in CCN would increase cloud albedo by about 0.03. If it is further assumed that marine cloud cover is the same as today (about 31% of the marine area) then the direct radiative effect of this increased albedo would cause an additional surface cooling of 1°C. Legrand *et al.* consider that the maximum cooling would have occurred over the period 70–15 ka BP.

During the course of the above discussion, we have identified a number of potential feedback effects involving changes in atmospheric composition over glacial–interglacial cycles. These range from highly plausible (CO_2) to highly speculative (ocean productivity – cloud albedo) mechanisms. In terms of helping to explain the links between orbital forcing and climate response, all the proposed feedbacks operate in the required direction (they are positive feedbacks amplifying the initial forcing) and on the required scale (global or hemispheric). On the basis of multivariate analyses, Lorius *et al.* (1990) conclude that three of these mechanisms (changes in the concentration of greenhouse gases, aerosol loading and marine *nss* sulphate), together with solar insolation and ice volume changes, are sufficient to explain over 90% of the variance in the Vostok temperature record.

2.4.3 The oceans

In order to explain the observed hemispheric synchroneity of glaciation, despite periods of directly opposed orbital forcing in the two hemispheres, many researchers have looked to the oceans. Before discussing a number of theories concerning the role of the oceans, we first outline some key features of the present-day ocean circulation system (Figure 2.9).

At present, northern maritime Europe is warmed by heat carried polewards by the Gulf Stream. The warm water meets cold air near Iceland; the air warms and the water cools and sinks. The bottom water so formed (North Atlantic Deep Water (NADW)) flows southward through the western Atlantic, round Southern Africa and Australia, and then northwards into the Pacific Ocean. The system is thought to be driven by salinity variations. The North Atlantic is warmer than the North Pacific. Evaporation is, therefore, greater in the North Atlantic and seawater is more saline than in the North Pacific. Net water vapour transport across the Panama Isthmus from the Atlantic to the Pacific Ocean leaves behind excess

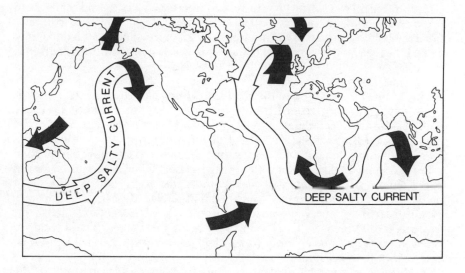

Figure 2.9 The present-day thermohaline circulation system. (After Broecker and Denton, 1989.)

salt, thus further increasing North Atlantic salinity. Under the present-day system, different circulation regimes dominate in the Northeast Atlantic and Northeast Pacific Oceans. These differences are reflected in the climates of the neighbouring coasts. The dominant surface current in the Northeast Atlantic is the warm, polewards flowing, Gulf Stream. In contrast, the dominant surface current in the Northeast Pacific Ocean is cold and flows towards the Equator.

Those components of the ocean circulation system which are driven by salinity variations are referred to as the *thermohaline circulation*. A number of the theories which have been put forward concerning the role of the oceans in the processes of glaciation and deglaciation invoke changes in the rate of NADW production and other characteristics of the thermohaline circulation.

It has been suggested that sea-level change provides a link between terrestrial and grounded marine ice sheets and so is capable of translating fluctuations in Northern Hemisphere ice sheets into a global climate signal (Denton and Hughes, 1983; Denton *et al.*, 1986). This mechanism has been invoked to explain the rapid nature of glacial terminations. At the onset of deglaciation, it is argued, increases in summer insolation and temperature near the margins of the Northern Hemisphere ice sheets cause substantial ice melt and recession and lead to the northward migration of the North Atlantic polar front. Formation of NADW, prevented during glacial periods, can then recommence. It is postulated that these changes may control atmospheric CO_2 concentration changes. Rising sea level and greenhouse warming both contribute to the melting and break-up of the grounded Antarctic ice shelf, so initiating Southern Hemisphere deglaciation

(Denton *et al.*, 1986). Thus, it is proposed, increased Northern Hemisphere summer insolation, rising sea level and increased CO_2 act together to initiate the irreversible collapse of the interlocked global ice-sheet system, at a time when it has reached its largest and most vulnerable configuration (Denton *et al.*, 1986).

Substantiation of this theory requires reliable evidence of early deglaciation at the margins of the Northern Hemisphere ice sheets. Analysis of an oxygen isotope record from the Norwegian/Greenland Sea directly dated by the accelerator mass spectrometry (AMS) method suggests that the marine-based Barents Shelf ice sheet disintegrated rapidly at about 15 ka BP. At the time, the dating of this event placed it considerably earlier than other recorded deglacial events following the Last Glacial Maximum (Jones and Keigwin, 1988). Uncertainty surrounds the extent and magnitude of melt, but Jones and Keigwin conclude that deglaciation in the Northern Hemisphere may have triggered the rapid collapse of the Antarctic ice sheet. More recent evidence suggests that retreat of the Fennoscandinavian ice sheet was also initiated at about 15 ka BP (Lehman *et al.*, 1991). This is some 2 ka earlier than previous estimated dates for this event.

Additional supporting evidence comes from an empirical study of the Antarctic ice sheet which indicates that melting occurred at a rate that was synchronous with, or lagged behind, the melting rate of the Arctic ice sheet (Nakada and Lambeck, 1988). Modelling experiments show that the Antarctic Peninsula ice sheet is particularly sensitive to sea-level change and to basal marine melting (Payne *et al.*, 1989). There is also evidence from a sensitivity analysis, carried out using a model of global sea-level change and almost 400 post-glacial sea-level records, which indicates that the West Antarctic ice sheet disintegrated in phase with the Northern Hemisphere ice sheets (Peltier, 1988a). While there is no evidence of Antarctic deglaciation occurring before initiation of Northern Hemisphere ice-sheet retreat, the available evidence does suggest that any linking mechanism must be rapid, if not instantaneous.

The proposed sea-level linkage is plausible so far as Antarctic deglaciation is concerned, but cannot account for changes in ice-sheet and glacier extent in inland areas (Broecker and Denton, 1990). It is also uncertain whether the proposed mechanism can play any role in glaciation and ice-sheet development. It is possible, although not supported by modelling evidence, that the development of Northern and Southern Hemisphere ice sheets may affect the climate of each hemisphere through ice–albedo effects rather than through a sea-level linkage, thus initiating mountain glaciation in both hemispheres (Manabe and Broccoli, 1985; Broecker and Denton, 1989).

The theory that Southern Hemisphere events may be initiated or controlled by events in the Northern Hemisphere appears to be contradicted by the analysis, by the SPECMAP group, of spatial variability, leads and lags in SST and $\delta^{18}O$ records from the Atlantic and Pacific Oceans (Imbrie *et al.*, 1989, discussed in Section 2.2). In this study, Southern Hemisphere SST was found to lead the climatic response to orbital forcing. In order to investigate potential mechanisms of change, Imbrie *et al.* also looked at the response patterns of $\delta^{13}C$ content and Cd/Ca ratios. These are both indicators of the efficiency with which nutrients are driven into the deep ocean.

The sequence of responses to both eccentricity and obliquity forcing was found to be identical: first Cd/Ca, followed by $\delta^{13}C$, then SST at 40°S and finally $\delta^{18}O$ (Imbrie et al., 1989). In the case of precessional forcing, Cd/Ca and SST respond first and at the same time. These sequences, it is concluded, imply that changes in the circulation and carbon chemistry of the deep ocean, together with changes in the surface ocean at high southern latitudes, are an early and essential part of the response to orbital forcing during interglacial transitions (Imbrie et al., 1989).

Finally, the SPECMAP group tested simple models of radiation forcing in the precessional and obliquity bands against the observed phase lags between $\delta^{18}O$ and SST. The best explanation of the observations was considered to be provided by a model in which summer radiation changes at high northern latitudes and at both orbital frequencies are transmitted to the Southern Hemisphere via changes in North Atlantic Deep Water production (Imbrie et al., 1989). The authors point out that these changes would occur at high northern latitudes beyond the range of the data used in the study (54°N) and very early on in each climate cycle. Detailed hypotheses are being developed and tested. While the SPECMAP group study supports the leading role of high northern latitudes in the response to orbital forcing, the proposed linking mechanism involves changes in deep-water production rather than sea-level change as suggested by Denton et al. (1986).

There is now substantial evidence of interglacial–glacial changes in ocean circulation, chemistry and vertical structure, although the interpretation of some of the geological indicators and estimated changes remains controversial. Most of the evidence relates to the last glacial–interglacial cycle and, in particular, to post-glacial events in the North Atlantic (the last deglaciation, Younger Dryas and Holocene). Major glacial and interglacial changes in North Atlantic ocean circulation and chemistry are summarized in Table 2.5.

The differences between glacial and interglacial periods and the present day summarized in Table 2.5 point to the mechanisms which may be involved in the processes of glaciation and deglaciation. During glacial episodes, for example, the lowest sea-surface temperatures and hence the smallest evaporation rates are found in the North Atlantic (Boyle and Keigwin, 1987). Deep-water production in the North Atlantic is reduced (Keffer et al., 1988; Keigwin and Jones, 1989). Changes in deep-water and intermediate-water production during the Last Glacial Maximum in the Atlantic, Indian, Pacific and Southern Oceans have been reviewed by Duplessy and colleagues (Duplessy et al., 1989; Duplessy and Juillet-Leclerc, 1991). The evidence suggests that most bottom- and deep-water production occurred in the high latitudes of the Southern Ocean, rather than in the North Atlantic as today. In the Atlantic, Indian and Pacific Oceans, intermediate waters extended somewhat deeper than at the present day. A well-defined and deep thermocline separated the intermediate and deep waters in the Indian and Pacific Oceans.

Table 2.5 Summary of major climate-related changes in the North Atlantic in glacial and interglacial episodes.

GLACIAL EPISODES (Last Glacial Maximum/Younger Dryas)

- lower sea-surface temperatures cause high regional surface pressure anomalies, increases in West Atlantic trade winds and reduced precipitation
- lower high-latitude sea-surface temperatures enhance intermediate-water production at expense of deep-water production
- polar front moves southwards
- evaporation is reduced
- deep-water production is reduced
- deep-water temperatures are lower
- salinity increases
- nutrient concentration of surface increases, but intermediate water becomes nutrient depleted
- ocean productivity increases
- ocean alkalinity increases

INTERGLACIAL EPISODES (Holocene)

- essentially all the glacial processes are 'reversed', in particular:
- deep-water production increases
- evaporation increases
- deep water is nutrient-depleted
- North Atlantic is warmer/more saline than the North Pacific
- sea-surface temperatures are higher
- northern Europe warms because of higher sea-surface temperatures, northward movement of Polar Front, establishment of Gulf Stream, etc.

Sources: Duplessy and Shackleton, 1985; Boyle and Keigwin, 1987; Broecker, 1987; Labeyrie *et al.*, 1987; Keffer *et al.*, 1988; Keigwin and Jones, 1989; Mix, 1989; Overpeck *et al.*, 1989.

It has been suggested that changes in the relative salinity content of the Atlantic and Pacific Oceans (caused by processes such as changes in sea-surface temperatures and evaporation rates, changes in freshwater and meltwater input and changes in water vapour transport) may cause the major ocean circulation patterns (Figure 2.9) to be reversed during glacial periods (Stommel, 1957; Gordon, 1982; 1986; Broecker, 1987; Broecker and Denton, 1989; 1990). The loss of northward-flowing warm surface-water masses in the Northeast Atlantic (the Gulf Stream) could cause northern Europe to cool by 6–8°C (Broecker, 1987).

Broecker's theory of mode changes in ocean circulation was developed in order to explain the Younger Dryas event (about 11 ka BP), but is also considered applicable to glacial periods. During this abrupt event, less than 1 ka in length, deglacial warming was halted and climatic conditions in the North Atlantic and surrounding areas deteriorated. The geographical extent of the event is, however, uncertain: the geological evidence is discussed in Chapter 3.

Broecker argues that the onset of the Younger Dryas corresponds with the major diversion, from the Gulf of Mexico to the Gulf of St Lawrence, of meltwater from the Laurentide ice sheet (Broecker and Denton, 1989). This diversion is attributed to a lobe of ice which formed as the Laurentide ice sheet retreated. The sudden influx of freshwater to the North Atlantic lowered salinity and deep-water production was reduced. The resulting circulation changes were reversed when the remaining ice lobe melted, diverting meltwater back to the Gulf of Mexico (Teller, 1990). In order to test fully this theory it is necessary to have well-dated, high-resolution chronologies of events. The relative timing of meltwater, sea-level and $\delta^{13}C$ peaks is crucial.

Broecker concedes that the shutdown of the North Atlantic 'heat conveyor system' (the Gulf Stream) alone is not sufficient to initiate global temperature changes and ice-sheet development (Broecker and Denton, 1989). He suggests that lower atmospheric concentrations of greenhouse gases (CO_2 and CH_4) and increased aerosol loading, together with reduced ocean heat transport, may all be necessary prerequisites for ice-sheet development. The length of glacial–interglacial cycles may then be set by the ice sheets themselves. Large ice sheets could, for example, alter the patterns of water vapour flux in the atmosphere and so affect the relative salinity of the Pacific and Atlantic Oceans. Broecker concludes that both large ice sheets and strong seasonal insolation cycles at high Northern Hemisphere latitudes are necessary to cause the ocean system to flip back into the interglacial mode of operation (Broecker and Denton, 1989). Broecker describes a non-linear system in which the atmosphere and ocean are linked together in such a way that the combined system is susceptible to abrupt mode changes. More than two modes may exist (Broecker and Denton, 1989).

Results from experiments with a coupled atmosphere–ocean GCM developed at the Geophysical Fluid Dynamics Laboratory in the USA provide some support for the suggestion that the coupled ocean–atmosphere system may have at least two stable modes of operation (Manabe and Stouffer, 1988). Two different stable equilibria are produced by the model, depending on which of two different initial conditions are specified. In the first case, a strong thermohaline circulation associated with a highly saline North Atlantic is specified, whereas the thermohaline circulation is absent in the second case. It is noted that the climates associated with each of these modes bear some resemblance to conditions during interglacial and the Younger Dryas/glacial events, respectively (Manabe and Stouffer, 1988). A number of shortcomings in the model are, however, apparent. The air–sea exchange of water, for example, has to be adjusted to remove a systematic bias in the model which suppresses the strength of the thermohaline circulation in the North Atlantic (Manabe and Stouffer, 1988). In addition, the model does not incorporate seasonal variations.

Further modelling evidence in support of thermohaline mode changes comes from Stocker and Wright (1991). They used an idealized ocean model to investigate whether or not small changes in the atmospheric flux (F) of freshwater from the Atlantic to the Pacific could force the thermohaline circulation to switch between modes. The model consists of two coupled, zonally averaged basins, representing the Atlantic and Pacific. They find that a decrease in F can lead to the reversal of the thermohaline circulation in the Atlantic (from mode A, the Atlantic conveyor belt circulation, to mode B, no Atlantic conveyor belt). No major changes occur, however, in the Pacific thermohaline circulation, in contradiction to Broecker's theory (Broecker and Denton, 1989).

Once in mode B, F can be restored to its present value without significant changes occurring in the thermohaline circulation. The system, therefore, has two stable modes under identical boundary conditions. To re-establish the Atlantic conveyor belt (mode A), F must be increased to 1.2 times its present value. These model results indicate that the likelihood of mode A–mode B transitions occurring depends critically on the initial state and direction of change of the hydrological cycle (i.e. on changes in F). In order to estimate the stability of the system at any given point in time, it is necessary to know how close it is to a transition point. The closer it is, the more sensitive it is to relatively small changes.

Simple box-model calculations have been used to investigate the almost instantaneous propagation of the meltwater signal across the entire North Atlantic during the last deglaciation (Duplessy et al., 1991). The model has four oceanic boxes (surface Atlantic, deep Atlantic, surface Indo-Pacific, and deep Indo-Pacific) and one well-mixed atmospheric box which is linked to the two surface ocean boxes. The ocean boxes are linked by a simple thermohaline circulation system. When meltwater is added to the surface Atlantic box, the signal reaches Pacific and Indian deep waters within less than 1 ka. The model results suggest that the strength of the thermohaline circulation (as measured by the flux of freshwater from the Atlantic to the Pacific) increased rapidly at the onset of deglaciation and then more gradually until 10 ka BP. It is argued that the speed of these changes provides additional support for mode changes related to the thermohaline circulation (Duplessy et al., 1991).

The evidence in support of mode changes is still controversial (Berger, 1990). However, Broecker is not alone in giving deep-ocean circulation and chemistry changes a major role in the processes linking orbital forcing and the global climate response. The SPECMAP group reach the same conclusion (Imbrie et al., 1989). New geological data indicate that coupled variations in North Atlantic surface temperatures and NADW may have occurred prior to the Last Glacial Maximum during oxygen isotope stage 3 (Boyle and Rosener, 1990). The very latest $\delta^{13}C$ record from the Southern Ocean 'provides strong evidence that NADW could have been the primary amplifier of glacial–interglacial climate change in high latitudes' (Charles and Fairbanks, 1992). There is also general agreement that changes in the atmospheric concentration of CO_2 are in some way controlled by changes in marine productivity and ocean structure.

2.4.4 Geographic boundary conditions

It is generally acknowledged that changes in continental configuration and altitude, related to continental drift and polar wandering, have played a role in the development of climate conditions during earlier periods of the Earth's history (Frakes, 1979). It is probable, for example, that continental drift created the prerequisite boundary conditions for the ancient Precambrian, Ordovician–Silurian and Permo-Carboniferous ice ages and for the ensuing Mesozoic climate optimum (see Figure 1.3). It is also likely that continental drift may have contributed towards global cooling during the Tertiary (about 65–2.5 Ma BP), through processes such as the separation of Antarctica and Australia.

There is now evidence to suggest that continuing tectonic uplift has produced at least some of the geographical boundary conditions necessary for the initiation of Northern Hemisphere glaciation. This evidence is published in a series of three papers (Kutzbach *et al*., 1989; Ruddiman and Kutzbach, 1989; Ruddiman *et al*., 1989), which are summarized by Ruddiman and Kutzbach (1991) and reviewed by Kerr (1989a).

The first of these three papers contains a summary of the geological evidence for continuing tectonic uplift in Asia and the American west (Ruddiman *et al*., 1989). It is concluded that the Tibetan plateau has risen by 4 km over the last 40 Ma and by 2 km over the last 10 Ma alone. Two-thirds of the uplift and southeastwards tilting of the Sierra Nevada is estimated to have occurred over the same 10 Ma period. Uplift in these areas, together with the Bolivian Andes and New Zealand Alps, is thought to be continuing.

In the second paper, results of experiments conducted with a version of the Community Climate Model of the National Center for Atmospheric Research are described (Kutzbach *et al*., 1989). In these experiments, Northern Hemisphere conditions in January and July are simulated, but no changes in the seasonal extent of sea ice or ice sheets are allowed to occur. In order to test the sensitivity of atmospheric circulation to uplift, the model is run with the major mountain ranges of Asia and western America at their present-day altitude (the control run), at half their present-day altitude and at zero altitude. The presence of high mountain ranges is found to have two major climatic effects. First, the presence of a major obstacle diverts air flow. Second, major mountain ranges generate large columns of moving air, ascending in summer and descending in winter. These vertical draughts themselves form obstacles to air flow.

The implications of these two effects for global and regional climate change are explored in the third paper (Ruddiman and Kutzbach, 1989). Specific aspects of regional and global climate changes simulated by the model as altitude increases are tested against palaeoclimate data from the last 10 Ma or so. Fourteen changes are investigated, 12 of which, including colder northern European winters and drier Mediterranean summers, are seen in both the model and data. The simulated atmospheric circulation changes in the North Atlantic are thought to favour the increased production of deep and intermediate water. It is concluded that continuing tectonic uplift provides a physically and geographically coherent explanation for some of the major patterns of regional climate change in the Northern Hemisphere over the last 10–15 Ma. The estimated uplift of the Asian and western American mountain ranges is not, however, sufficient to explain all the observed cooling. Over the highest mountain plateaus, model January temperatures cool by up to 8°C. Summer temperatures cool by up to 2°C. This is still about 5–10°C less than the observed cooling during Northern Hemisphere glaciations. The introduction of additional variables to the model, such as varying snow cover, SST, soil moisture and atmospheric CO_2 may, Ruddiman and Kutzbach believe, provide the additional cooling.

There is widespread evidence of a major change towards severe climate conditions in the Northern Hemisphere at about 2.5 Ma BP (see Section 2.2 and Chapter 7). The modelling experiments discussed above suggest that continuing tectonic uplift has contributed towards this deterioration and has created geographical boundary conditions conducive to Northern Hemisphere glaciation. In Section 2.3 we described the model developed by Maasch and Saltzman (1990) which was able to simulate the emergence of the 100 ka cycle, if model parameters were gradually varied in order to represent some (unknown) type of slow tectonic change. It appears that this experiment may have some physical justification. It has, however, been argued that some of the evidence put forward in support of recent tectonic uplift by Ruddiman *et al.* (1989) has been misinterpreted (Molnar and England, 1990). The rate of river incision, for example, is used by Ruddiman *et al.* (1989) to infer uplift, whereas climate change can itself cause changes in erosion rates. Similarly, Molnar and England point out that vegetation changes can be attributed to climate change as well as to changes in altitude.

While the rate and timing of tectonic changes may be disputed, the influence of topography on certain features of the atmospheric circulation, such as the Indian monsoon, is widely accepted. It has also been suggested, but not yet proven or disproven, that the expansion of sensitive high-altitude Tibetan glaciers may have enhanced glacial cooling in other parts of the world through positive ice–albedo feedback effects (Kuhle, 1987).

Once glaciation is under way, the ice sheets themselves are considered to be an important component of the geographic boundary conditions. There is considerable evidence from GCM modelling experiments to show that large terrestrial ice sheets influence local and regional climate through their physical and mechanical topographic effects, as well as through their thermal and radiative effects (Kutzbach and Guetter, 1986; Mitchell *et al.*, 1988; Rind *et al.*, 1989; Shinn and Barron, 1989). This evidence is discussed in detail in Section 8.2. Ice-sheet elevation and configuration are thought to be important parameters so far as atmospheric circulation changes are concerned.

In terms of orbital forcing cause-and-effect mechanisms, it is likely that the slow response times of large terrestrial ice sheets play a more important role than do the circulation changes related to their topographic and thermal effects. It has, for example, been suggested that when ice sheets reach a critical size, the characteristic time of associated isostatic rebound starts to interact with some of the orbital frequencies, so producing combination tone frequencies (Berger, 1989b). It is speculated that this mechanism could account for the emergence of the 100 ka cycle in association with a trend towards more severe glacial episodes over the last 500 ka or so.

While mechanisms involving large ice sheets can be invoked to explain the orbital periodicities observed in the Quaternary record, they cannot be invoked for the pre-Quaternary. As described in Section 2.2, the Cretaceous, for example, was very different. It was characterized by global temperatures up to 12°C warmer than the present day and by a higher atmospheric concentration of CO_2. Geographic boundary conditions, such as continental configuration, were also very different. The continental area was less extensive because of high global sea levels. It is debatable whether or not polar ice sheets were present. Modelling studies suggest that these boundary conditions (increased CO_2, a different continental configuration and reduced continental area) all contributed towards warming and that these factors were amplified by enhanced poleward heat transport by the oceans (Barron, 1988). The observed sedimentary cycles (see Section 2.2) may then be explained by periodic changes in insolation, evaporation and precipitation (see Berger, 1989b, for a summary).

2.5 Conclusions

In this chapter, we have reviewed recent theoretical, modelling and empirical evidence in support of the Milankovitch theory of climate change. It sometimes appears that more questions have been raised than answered as the data quality has improved, and the distribution and type of available geological information has expanded. Certainly, it is now clear that orbital forcing, the climate response and the linking cause-and-effect mechanisms are complex, involving all components of the climate system (cryosphere, lithosphere, oceans, atmosphere and biosphere). The evidence indicates that both external and internal forcing are required to produce 100 ka and other orbital-related periodicities in the geological record. It has not yet been established whether orbital forcing acts as a driving or as a pacing mechanism of internal climate variability.

We have discussed a number of potential mechanisms linking cause-and-effect, including isostatic loading and unloading of the Earth's crust; changes in the atmospheric concentration of CO_2 and other greenhouse gases; changes in the atmospheric loading of terrestrial and marine aerosols; changes in cloud reflectivity; ice–albedo feedbacks; sea–level change; changes in ocean circulation, chemistry and vertical structure; and geographic boundary conditions. It is likely that a combination of many or all of these mechanisms has operated over the Quaternary.

This conclusion is supported by a number of modelling experiments (Adem, 1988; Harvey, 1989a; 1989b; Hyde et al., 1989; Rind et al., 1989). Adem (1988), for example, used a thermodynamic model to simulate the effect of changes in insolation, atmospheric CO_2 and surface–albedo feedbacks on Northern Hemisphere climate. The model underestimates cooling at 18 ka BP and Adem concludes that some feedback mechanism(s) may be missing. On the basis of GCM and EBM modelling experiments, Hyde et al. (1989) conclude that variations in deep-ocean heat transport may account for, at maximum, only 20–30% of Antarctic temperature variability over the last interglacial–glacial cycle. Rind *et al.* (1989) tried to initiate ice-sheet development in the Goddard Institute for Space Studies GCM. Even with insolation changes, lower CO_2, increased sea ice and lower SST, the model had problems maintaining the ice sheets. All these results indicate that at least one of the mechanisms responsible for glacial cycles is still missing from our equations.

Despite the remaining uncertainties, we conclude that orbital changes, as described by the modified Milankovitch theory, are the major cause of the glacial–interglacial cycles observed over the Quaternary. However, we stress that orbital forcing can only ever explain part of the variability of the palaeoclimate record. Climate changes over many different time scales. The spectrum of climate variability contains a large component of random forcing or noise (Mitchell, 1976). Part of the variability of Quaternary climate can be attributed to shorter-term forcing and to mechanisms of regional rather than global climate change. These are discussed in the next chapter.

3 Short-term climate change

3.1 Introduction

Climate changes occurring over time scales shorter than those associated with the orbital forcing frequencies are defined here as short-term (see Section 1.2). Fluctuations on time scales less than 100 a are considered here as climate variability, and are not discussed. The shortest period which has been identified in the orbital data is 15 ka (Berger *et al.*, 1989a). Spectral analysis of the geological record indicates that orbital forcing is concentrated in the frequency band 20–100 ka (Shackleton and Imbrie, 1990). This is not to say that orbital forcing has no effect over time scales less than about 20 ka.

In Figure 3.1, we show the boundary conditions used by the Cooperative Holocene Mapping Project (COHMAP) group for GCM snapshot simulations of global climate over the last 18 ka (COHMAP, 1988). The seasonal distribution of Northern Hemisphere radiation varies strongly over this post-glacial period (Berger, 1978a). At 18 ka BP, the seasonal and latitudinal distributions of insolation were similar to those of today. As the Earth–Sun distance decreased in the northern summer, and as obliquity increased, the seasonal contrast in radiation receipts became more pronounced in the Northern Hemisphere and less pronounced in the Southern Hemisphere. Maximum Northern Hemisphere seasonality was reached at about 9 ka BP, when Northern Hemisphere radiation was, on average, 8% greater in July than at the present day and 8% less in January. Since 9 ka BP, the strength of the Northern Hemisphere seasonal radiation cycle has reduced. Using these insolation changes with the internal boundary conditions shown in Figure 3.1, the Community Climate Model developed by the National Center for Atmospheric Research is able to reproduce many of the known major features of the palaeoclimate (see discussion in Section 8.2).

Orbital variations can be shown to have some influence on climate over even shorter time scales. Adem (1989) has investigated the effect of computed insolation distributions at 4 ka and 2 ka BP (Berger, 1978b) on Northern Hemisphere climate as simulated by a thermodynamic climate model. The insolation patterns of 4 ka and 2 ka BP produce stronger seasonal temperature cycles than those of the present day, particularly at low latitudes, although average annual temperatures at each latitude were virtually identical to those of the present day. At 4 ka BP, mean spring temperatures of continental interiors were 1.0°C lower than the present day, whereas summer temperatures were 1.1°C greater (Adem, 1989). The changes at 2 ka BP were approximately half those found at 4 ka BP. Due to the thermal inertia of the oceans, the changes in SSTs are much smaller.

Spectral peaks in the geological record may be related to orbital forcing even though their frequency is less than that of the shortest computed orbital

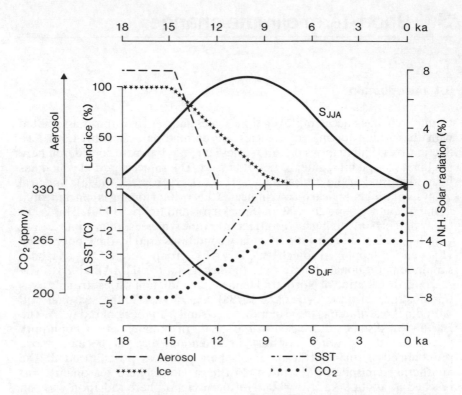

Figure 3.1 Boundary conditions used by COHMAP (1988) for their GCM palaeoclimate snapshot simulations of the last 18 ka. (From Kutzbach and Street-Perrott, 1985).

S_{JJA} = Northern Hemisphere solar radiation June–August (percent departure from present day)

S_{DJF} = Northern Hemisphere solar radiation December–February (percent departure from present day)

Ice = land ice (percent of 18 ka BP ice volume)

SST = global mean sea-surface temperature departure from present-day (°C)

CO_2 = atmospheric CO_2 concentration (ppmv)

Aerosol = excess glacial-age aerosol (arbitrary scale).

frequencies (i.e. 15 ka). For example, analysis of three $\delta^{18}O$ records from the Indian Ocean, which have a sampling resolution of 500 a, reveals spectral maxima at 10.3 ka, 4.7 ka and 2.5 ka (Pestiaux *et al.*, 1988). These maxima, together with smaller spectral peaks at 13.0 ka, 10.4 ka and 9.4 ka, are close to the estimated combination tones of the obliquity and precessional cycles. It is concluded that the observed spectrum is in general agreement with the expected response of a non-linear climate oscillator forced by insolation variations (Le Treut and Ghil, 1983).

We have defined short-term climate change as occurring over time scales from 100 a to 20 ka (Section 1.2). A number of major mechanisms operating over these time scales, apart from orbital forcing, can be identified (Wigley, 1981; Goodess *et al.*, 1988). These are changes in ocean circulation, evolution of the atmosphere, volcanic activity, air–sea–ice–land feedbacks, solar variability and atmosphere–ocean feedbacks. Some of the mechanisms are periodic and potentially predictable (for example, solar variability), while others are essentially random and are not, therefore, predictable (volcanic activity). Some mechanisms are external to the climate system (solar variability), whereas others are internal (changes in the composition of the atmosphere). In Section 3.3 we assess these potential causal mechanisms. First, however, we describe the pattern of climate change over the last 20 ka in the western European and North Atlantic region.

3.2 Geological evidence of post-glacial climate change

The principal sources of information on climates of the last 20 ka are high-resolution ocean sediment cores, ice cores, pollen records, fossil beetle remains, glacier fluctuations, tree rings, fluctuating snowlines and treelines, and historical records (Huntley and Birks, 1983; Williams and Wigley, 1983; Bradley, 1985; Röthlisberger, 1986; Grove, 1988). The record of climate change over the British Isles is discussed in detail in Chapter 7. Here we are concerned with general patterns of climate change and variability.

We have already established that the major features of the record of the last 20 ka can be linked to orbital forcing. These features can conveniently be divided into three principal periods: the Last Glacial Maximum (18 ka BP); the warming period (about 15–10 ka BP); and the Holocene warm period (10 ka BP to date). Superimposed on these major periods are shorter-term climate fluctuations. In this section, we describe some of these fluctuations and evaluate their potential causes. The Younger Dryas marked a sudden reversion to cooler conditions at the end of the warming period. Throughout the Holocene, climatic fluctuations have occurred, marked by a range of proxy indicators such as the position of treelines and glacier advances and retreats. One of these climate events, the Little Ice Age, is considered separately here because of the wealth of information that has become available in recent years.

3.2.1 The Younger Dryas

There is widespread evidence that a period of rapid climate deterioration occurred about 11 ka BP. This event persisted for a few hundred years at most, and is known as the Younger Dryas. The Younger Dryas is associated with the readvance of glacial ice in Scotland and the Lake District (the Loch Lomond Advance) and with a widespread return to cooler, more severe climatic conditions over the British Isles (see Chapter 7). This is the last time that glacial ice was present in England and Wales.

Until recently, the Younger Dryas event was considered to be restricted to northwestern Europe. However, in recent years, evidence for its existence has emerged from Greenland, eastern Canada, the North Atlantic as far south as 28°N, mid-continental USA, southern South America, Antarctica, the North Pacific, northeastern USA and the South China Sea region (Broecker *et al.*, 1988; Keigwin and Jones, 1989; Peteet *et al.*, 1990; Kudrass *et al.*, 1991). There is also some evidence of a significant dry event in the tropical region of northern Australia at approximately the same time, which has been correlated with the Younger Dryas (De Deckker *et al.*, 1991). Despite these findings, the Younger Dryas cannot be described as a global event. The most convincing evidence is concentrated in the North Atlantic and surrounding areas, which suggests that the cause is linked to changes in atmospheric and/or ocean circulation over the North Atlantic Ocean (see Section 3.3.1).

One of the most striking features of the Younger Dryas is its abrupt start and finish. Analysis of the isotope and dust content of Swiss lake sediments and Greenland ice cores suggests that it ended abruptly at about 10.7 ka BP (Dansgaard *et al.*, 1989). It is estimated that in less than 20 a the climate of the North Atlantic returned to milder and less stormy conditions and that sea ice retreated rapidly. Southern Greenland is thought to have warmed by 7°C in about 50 a and precipitation may have increased by 50% over the same period (Dansgaard *et al.*, 1989).

3.2.2 Climate fluctuations within the Holocene

The Holocene (10 ka BP to present) was once thought of as a period of consistently warm conditions. This view has been modified by emerging evidence which reveals it as a period of oscillating climate. Denton and Karlén (1973) identified three global glacial readvances, at 5.3 ka, 2.8 ka and 350–250 a BP. Many more episodes of glacial readvance and retreat have since been demonstrated, from studies of pollen cores, glacial moraines, lichen, treelines and tree rings.

Evidence from the Alps, Scandinavia, Iceland, Asia, North America, South America, New Zealand and equatorial glaciers has recently been reviewed by Grove (1988). She concludes that, in regions where the data are most reliable, a number of common features can be identified in Holocene chronologies. Periods of glacial advance and retreat are more or less equal in length. The frequency of these fluctuations increases towards the present day, but this may be a reflection of improvements in the resolution and reliability of data. Grove notes that the most recent advances tend to occur at intervals of about 1 ka and cites evidence in support of globally near-synchronous advances at about 4.3 ka, 3.2 ka and 2.5 ka BP.

Röthlisberger (1986) analysed glacial moraine sequences collected from Northern and Southern Hemisphere glaciers of similar size, using a consistent sampling methodology. Radiocarbon dating of all samples was carried out in the same laboratory. This internally consistent data set was then used to assess glacier fluctuations in seven regions of the world (Figure 3.2). The reliability of the regional time series varies, but those from New

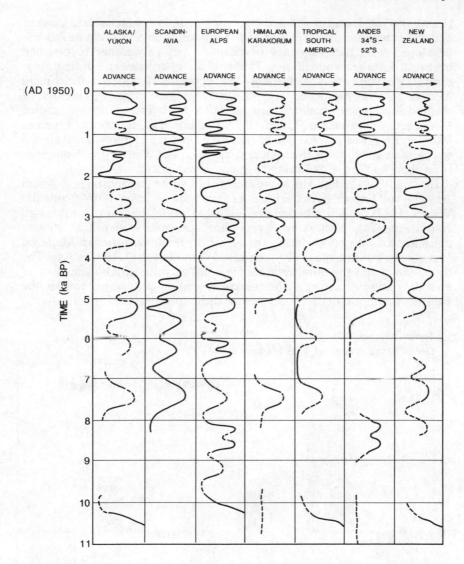

Figure 3.2 Fluctuations of glaciers in the Northern and Southern Hemispheres in the Holocene (10–0 ka BP). (From Röthlisberger, 1986.)

Zealand and the Alps are considered to be the most reliable. The good agreement between these two series, it has been argued, suggests that globally synchronous phases of glacial expansion and retreat have occurred throughout the Holocene (Röthlisberger, 1986; Grove, 1988).

A feel for the magnitude of climate variability experienced in the British Isles over the Holocene is given by estimates of temperature and precipitation in England and Wales extrapolated from historical records

55

(Lamb, 1977a). Lamb developed regression equations which relate summer and winter temperature averages and precipitation totals to climate indexes developed from the frequency of various types of weather reported in historical documents and diaries. These equations are used to estimate long-term temperature averages for individual episodes identified in the climate indexes for the Holocene (Lamb, 1977a; Figure 3.3). A number of methodological shortcomings are acknowledged, including the lack of independent verification (Lamb, 1977a). Caution is, therefore, urged in interpreting these results. However, they do suggest that mean annual temperatures may have fluctuated by as much as 3°C over the Holocene, from about –1°C to +2°C around present-day values.

In a study of the late Holocene, records of summer conditions in North America and Europe over the last 2 ka have been reviewed by Williams and Wigley (1983). Pollen, treeline, tree-ring and glacial deposit records are all considered. The authors recognize three previously identified climate episodes: a cold period during the eighth to tenth centuries; a Medieval Warm Epoch during the twelfth century, followed by the Little Ice Age. The exact timing and the number of minima and maxima within each of these episodes, however, varies from region to region. We now consider the evidence in support of the most recent of these episodes, the Little Ice Age.

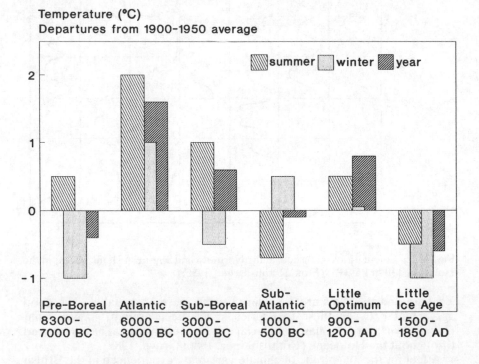

Figure 3.3 Estimated England and Wales temperature for selected periods during the last 10 ka. (Data from Lamb, 1977a.)

3.2.3 The Little Ice Age

A wide variety of dates has been ascribed to the Little Ice Age. In Europe, it is generally defined as beginning in the thirteenth or fourteenth centuries and culminating somewhere between the mid-sixteenth and the mid-nineteenth century (1550–1850).

The view that the Little Ice Age was a global event and not, as previously thought, a regional event confined to Europe and North America is gaining popularity. Grove (1988) has reviewed evidence in support of the Little Ice Age from Iceland, Scandinavia, the Alps, Asia, North America, Greenland, South America, New Zealand and Antarctica. The readvance of mountain glaciers, lowering of treelines, sea-ice expansion, increased erosion and flooding, agricultural decline, population decline and the abandonment of settlements (principally in Greenland and Iceland) are all cited as evidence of climatic deterioration. However, the influence of political and economic factors on human activity, together with events such as the Black Death epidemic in Europe, means that agricultural and social changes cannot be ascribed solely to climate change. Grove concludes that, for large parts of the globe, summer temperatures were 1–2°C cooler than today over several hundred years during the Little Ice Age. Interannual variability may have been greater and the Little Ice Age itself may have consisted of a series of relatively short climate fluctuations.

The available evidence reveals regional variations in the onset, maximum, termination and other phases of the Little Ice Age (Williams and Wigley, 1983; Grove, 1988). During the late eighteenth century, for example, glaciers were retreating in Scandinavia but advancing in the Alps.

Interpretation of the evidence from glacier movement is complicated by the wide regional variation in the availability of measurements. The first observations of North American glacier movements (from Alaska) were not made until the late eighteenth century. The longest available records come from the Swiss Alps and date from the sixteenth century. Detailed observations of the Rhonegletscher glacier have been made since 1546. Very few data are available from glaciers at low latitudes and in the Southern Hemisphere. A further problem is that the Little Ice Age cannot always be clearly identified in proxy climate indicators. In a 1.4 ka tree-ring record of summer temperature from Fennoscandinavia it is expressed only as a short period (c. AD 1570–1650) of limited cooling (about –0.5°C) (Briffa et al., 1990).

3.3 Causes of short-term climate change

While it is generally accepted that orbital forcing is responsible for the 'broad sweep' of climate change over the post-glacial period, the causes of shorter-term climate variability over this period are still uncertain. It is, however, likely that a number of different forcing mechanisms are involved. It has been suggested that the Younger Dryas was linked to changes in ocean circulation (Broecker et al., 1985). Climate variability in the Holocene can be linked to solar variability, major volcanic eruptions and to internal variability involving ocean-atmosphere/ice–albedo feedback effects (Grove, 1988; Wigley and Kelly, 1990).

3.3.1 Ocean circulation

Convincing evidence of the Younger Dryas event is concentrated in the North Atlantic region (see Section 3.2.1), which suggests that changes in the heat budget of the high-latitude North Atlantic Ocean may have played a major causal role. In Section 2.4.3, we introduced the theory put forward by Broecker and colleagues (Broecker and Denton, 1989) concerning mode changes in ocean circulation. They suggest that the onset of the Younger Dryas corresponds with the major diversion, from the Gulf of Mexico to the Gulf of St Lawrence, of meltwater from the Laurentide ice sheet. A number of authors have reasoned that, following the early stages of deglaciation, the southern exit from the meltwater Lake Agassiz was temporarily blocked by a lobe of the retreating ice sheet and that the outflow was diverted down the Gulf of St Lawrence (Teller, 1987; 1990; Street-Perrott and Perrott, 1990). This sudden influx of freshwater to the North Atlantic would lower salinity and hence reduce deep-water production. The present-day circulation system would be reversed (Figure 2.9) and the northward flowing 'heat conveyor' temporarily cut off, thus cooling western Europe.

If this theory is correct, then a reduction in North Atlantic Deep Water production should be seen just before the onset of the Younger Dryas, together with decreased salinity in the northwest North Atlantic. The initial condition would have meltwater draining into the Gulf of Mexico. In support of this, Keigwin and Jones (1989) have found a reduction of $\delta^{18}O$ in planktonic foraminifera from the Bermuda rise at about 13.5 ka and 12 ka BP. This indicates lower salinity, and can probably be associated with the discharge of meltwater from the Laurentide ice sheet into the Gulf of Mexico prior to the Younger Dryas.

Boyle and Keigwin (1987) argue that reduced $\delta^{13}C$ and increased Cd/Ca ratios in North Atlantic sediments can be dated to the onset of the Younger Dryas and are both indicative of a reduction in the production of North Atlantic Deep Water. Stoker *et al.* (1989) have found a marked increase in sedimentation and bioturbation, indicating increased bottom-water activity, in the Rockall Trough and Faeroe-Shetland Channel (Northeast Atlantic Ocean) at the end of the Younger Dryas. They believe this is related to the restoration of the overflow of Norwegian Sea Deep Water into the North Atlantic, and hence to the restoration of North Atlantic Deep Water production.

The precise timing and sequence of events are crucial as far as this theory is concerned. If the estimated dates of events vary by only a few hundred years then very different conclusions may be reached. Revised dates of high sea-level stands have recently been obtained from the Barbados coral reef record (Fairbanks, 1989). According to the revised series, sea-level rise, and hence ice-sheet melt, almost stopped during the Younger Dryas period, suggesting that glacial meltwater could not have played a major causal role (Fairbanks, 1989; Shackleton, 1989). Berger (1990) argues that NADW production was reduced before and after, and not, as suggested by Broecker, during the Younger Dryas. However, modelling evidence suggests that once a circulation mode, similar to that which Broecker associates with the Younger Dryas, is established, it can be maintained with very little freshwater input (Manabe and Stouffer, 1988; Rooth, 1990).

The evidence for deep-water changes during the Younger Dryas is uncertain. Jansen and Veum (1990), for example, obtain somewhat different dates for events than do Boyle and Keigwin (1987). They conclude that there is no evidence from the δ^{13}C record supporting a reduction of deep-water production during the Younger Dryas. It is even possible, they suggest, that production may have been enhanced, due to changes in the wind field along the marginal ice zone and to surface salinity changes.

Jansen and Veum (1990) argue that the Younger Dryas should not be seen as an anomalous event. Rather, it should be seen as part of the natural, two-stage, process of deglaciation. The extreme conditions of the Younger Dryas could have been caused by shifts in the atmospheric circulation related to changing albedo feedback effects as the ice sheets disintegrated.

High-resolution records are not available for previous deglaciations. It is not, therefore, possible to ascertain whether events of a similar magnitude to the Younger Dryas have occurred during previous terminations. If such events could be identified, then it might be easier to establish their causes. In the meantime, the role of ocean circulation changes during the Younger Dryas remains controversial. Indeed, it has been suggested that a search for any one specific cause of this event may be futile because 'in a chaotic system near its point of bifurcation, small disturbances can result in large effects, from positive feedback amplification. Every link in the feedback loop is both cause and effect' (Berger, 1990). Berger considers that meltwater spikes, ice-sheet collapse, and/or volcanic eruptions could all have acted as trigger events to the feedback system during the Younger Dryas. He regards the Younger Dryas as a throwback to glacial conditions from an unstable deglaciating condition.

Reconstructions based on instrumental and proxy data suggest that ocean circulation changes may also have occurred during the Little Ice Age (Lamb, 1979; Grove, 1988). Lamb suggests that less northward transport occurred during the period 1790–1829, leading to lower SSTs in the northeast North Atlantic, North Sea and Norwegian Sea. During the earlier period, 1675–1704, he suggests that more northward transport may have occurred, leading to higher SSTs and resulting in sufficient moisture being available for ice-sheet build-up in Labrador and Greenland. Others have postulated that all the episodes of glacial readvance observed over the Holocene may have been accompanied by ocean circulation mode changes similar to those proposed by Broecker (Bjerknes, 1968; Weyl, 1968). However, no explanation of why such oscillations should occur at intervals of a few thousand years has been put forward (Grove, 1988).

3.3.2 Solar variability

Solar variability remains a controversial mechanism of climate change. Much of the evidence put forward in support of a solar–climate link has come from studies based on poor statistical methodologies and on poor or short data records. Also, perhaps most crucially, no realistic mechanism has been put forward to explain exactly how the effects of solar variability could be transferred to the Earth's surface and thus affect climate.

We now examine the role of solar variability, discuss some of the more recent empirical and modelling evidence and assess whether or not this mechanism could explain the climate variability observed over the Holocene. These issues have been reviewed elsewhere (Wigley, 1988a; Foukal, 1990) and have also been the subject of two major recent conferences (Stephenson and Wolfendale, 1988; Royal Society, 1990).

The most visible feature of solar variability, and that which has attracted the greatest attention, is the periodic variation in the number of *sunspots*. Observations of sunspot numbers were first made by telescope in about 1610, but reliable records are not available prior to 1848 (Foukal, 1990). Over the last 150 a, the length of sunspot cycles has been 10–12 a, averaging about 11 a. The amplitude of the cycles shows considerable variation. The peak of each cycle has ranged from only 45 sunspots (in 1804 and 1818) to 190 (in 1957).

Sunspot cycles are thought to be related to solar magnetic variations, and a double magnetic cycle (about 22 a) can also be identified. It is possible that the solar magnetic cycle is driven by the differential rotation of the Sun, but the mechanisms involved are poorly understood (Pittock, 1978; Foukal, 1990; Gough, 1990; Weiss, 1990). From the climatological point of view, however, the amplitude and time scales of the associated changes in solar irradiance (the solar constant) are more important.

It is considered that the direct climate response to a 0.1% change in solar variability over an 11-year sunspot cycle would be so small (about 0.03°C) as to be undetectable (Wigley, 1988a). Nevertheless, a number of authors claim to have identified 11- or 22-year solar cycles in the climate record.

One of the early, more statistically reliable, studies is that of Mitchell *et al.* (1979) who demonstrated a relationship between drought in the US High Plains and the 22-year magnetic cycle minimum. More recently Padmanabhan and Rao (1990) have found a significant relationship between the 22-year magnetic cycle, represented by the Zurich sunspot record, and drought indices in Iowa and Kansas. The relationship is, however, weak, explaining only 10–15% of the variance in the drought record. Twenty-one-year peaks in the global and hemispheric record of marine air temperature for the period 1856–1986 have been identified and are thought to show a negative correlation with the Zurich sunspot record (Newell *et al.*, 1989). Correlations have also been found between 11-year cycles in the Zurich sunspot record, the record of total atmospheric ozone from Arosa, Switzerland, and records of maximum and minimum temperature from Genoa, Italy (Mazzarella and Palumbo, 1989). Maximum production and minimum loss of ozone lag each sunspot maximum by 2.4 a. Variations in minimum temperature occur in phase with the ozone variations, while variations in maximum temperature lag the ozone variations by 5.8 a.

A series of papers which has recently attracted much attention suggests that the sunspot–climate link may be masked by interactions with the Quasi-Biennial Oscillation (QBO) (Labitzke and van Loon, 1988; 1989a; 1989b; 1990; van Loon and Labitzke, 1988). The QBO is the periodic reversal of stratospheric winds over the equator which occurs about every 13 months. Labitzke and van Loon have found significant and robust correlations between 11-year cycles in solar variability and climate records, provided that the climate data are first separated according to the phase of the QBO (Figure 3.4). Similar results have been found for a variety of Northern Hemisphere climate records, notably temperature and geopotential heights, both in winter and in summer, and for the troposphere and stratosphere. The correlations are found to be statistically significant and have stood repeated testing (Kerr, 1988). Comparable relationships have been established for other climate parameters: the variability of North Atlantic storm tracks (Tinsley, 1988), and Northern Hemisphere tropospheric temperature and wind (Venne and Dartt, 1990).

The data sets used by Labitzke and van Loon are relatively short, covering a maximum of four sunspot cycles. The analysis has been extended back to 1884 using SST records (Barnett, 1989). In this study, the climate–sunspot relationship is found to break down before about 1925. This may be attributable to poor data or to weaker solar forcing in the earlier part of the record. Or it may be that the postulated relationship is spurious (Barnett, 1989).

It is possible that the approximate 11-year cycle identified in so many climate records is caused by some internal oscillation and not by external

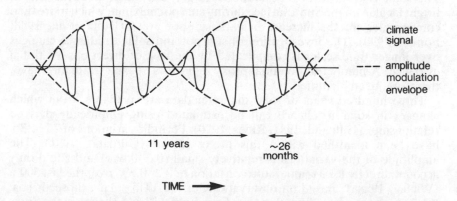

climate signal
--- amplitude modulation envelope

11 years   ~26 months

TIME ➡

Figure 3.4 Schematic representation of the solar cycle/QBO calculations carried out by Labitzke and van Loon (1989a; 1989b). The amplitude modulation envelope of the climate record is formed by the west and east phase of the Quasi-Biennial Oscillation (QBO). Thus, by sampling according to the west phase of the QBO (the crests of each QBO cycle) or by the east phase of the QBO (the troughs of each QBO cycle) 11-year cycles are produced. (After Barnett, 1989.)

solar forcing. A primitive-equation model of the global troposphere is, for example, able to generate an 11-year cycle from the internal non-linear dynamics of atmospheric flow (James and James, 1989). It is conceivable that, simply by chance, the phase of such an oscillation could coincide with the phase of solar variability, thus explaining results such as those of Labitzke and van Loon (Geller, 1989; James and James, 1989). Since no plausible physical mechanisms have been put forward to explain how sunspot cycles could cause the observed cyclical variations in climate, the postulated links between climate and 11- or 22-year sunspot cycles must remain controversial, despite the strength of some of the statistical relationships.

It has been suggested that long-term variations in the *amplitude* of sunspot cycles may have an influence on climate (Sofia *et al.*, 1979; Eddy, 1982). Sunspot records, based on observations made with the naked eye rather than by telescope, reveal a number of periods of either no, or very limited, sunspot activity, the best known of which is the Maunder Minimum (1645–1715). Other notable events include the Spörer Minimum (1450–1534) and Wolf Minimum (1282–1342). It has been suggested that the association of two of these events with the Little Ice Age (1350–1850) implies some causitive link (Eddy, 1976; 1977; Grove, 1988).

A few years of *solar irradiance* data recorded by satellite are now available. These data indicate that, on time scales of days to months, solar irradiance varies with sunspot number. Early studies suggested that solar irradiance decreases as the number of sunspots increases (Hoyt and Eddy, 1982). More recent data from the Solar Maximum Mission and Nimbus 7 satellites show that the solar constant decreased by about 0.1% from the peak of sunspot activity in 1981 to a minimum in mid-1986 (Foukal, 1990). This somewhat surprising observation is largely attributed to the increase in bright faculae on the Sun's surface during sunspot maxima, which more than compensates for the increase in dark sunspot areas (Kuhn *et al.*, 1988; Foukal, 1990). The investigation of sunspot and solar-irradiance changes over longer time scales is complicated by the lack of direct observational evidence. A number of potential proxy indicators of solar irradiance have, therefore, been identified.

Three hundred years of solar diameter data are available, from which changes in solar irradiance can be estimated using empirically derived relationships (Gilliland, 1981; Ribes, 1990). Periodic variations of 76 ± 8 a have been identified within this proxy record (Gilliland, 1981). The amplitude of the variations is relatively small (0.3–0.4%) and could only account directly for a temperature variation of 0.2–0.3°C over the last 300 a (Wigley, 1988a). Based on observations of the 1715 solar eclipse, it has, however, been argued that there has been no significant change in the Sun's diameter since 1715 (Morrison *et al.*, 1988).

The record of ^{14}C values in tree rings is potentially the most valuable proxy indicator of solar-irradiance changes. Changes in the output of energetic particles from the Sun (the solar wind) are believed to modulate the production of ^{14}C in the upper atmosphere (Stuiver and Quay, 1980; Stuiver *et al.*, 1991). The magnetic properties of the solar wind change over the

11-year solar cycle, and may in turn lead to variations of up to 20% in annual atmospheric ^{14}C production (Stuiver *et al.*, 1991). The effect of the solar wind is such that high ^{14}C production is associated with periods of low solar activity, i.e. periods of low sunspot numbers.

Variations in atmospheric ^{14}C are, in turn, thought to be correlated with variations of ^{14}C recorded in tree rings, after a long-term trend related to changes in the Earth's magnetic field has been removed. Changes in ocean circulation and/or global wind speeds are found not to affect the δ^{14}C record, although some adjustments to the record may be required for the Younger Dryas period (Stuiver *et al.*, 1991). As noted above, the proposed solar variability–δ^{14}C relationship is negative: periods of sunspot minima, such as the Maunder Minimum, coincide with high ^{14}C values in tree rings and vice versa (Eddy, 1976, 1977; Stuiver and Quay, 1980)

Eddy (1977) identified 18 episodes characterized by very low or high sunspot activity in the δ^{14}C tree-ring record of the last 7.5 ka. In terms of sunspot numbers, these episodes are comparable to the Maunder and Spörer Minima, and to the Medieval Maxima (AD 1120–1280), respectively. He then compared the δ^{14}C record with four estimates of past climate: Alpine glacier fluctuations (Le Roy Ladurie, 1967), global glacier fluctuations (Denton and Karlén, 1973); mean annual temperature in England (Lamb, 1977a); and a winter-severity index for the Paris–London area (Lamb, 1975). In each episode of low sunspot activity (high ^{14}C), mid-latitude glaciers advanced and temperatures fell (Eddy, 1977). In each episode of high sunspot activity (low ^{14}C), the glaciers retreated and temperatures rose. Eddy proposes that the changes in solar activity and climate may be linked by slow changes in the solar constant, of about 1% magnitude.

Relatively long and reliable records of δ^{14}C tree-ring values are now available. A number of spectral peaks have been identified in these records which are thought to be related to solar variability (Hood and Jirikowic, 1990; Sonett and Finney, 1990). Periodicities of 2.4 ka, 1.0 ka, 200 a and 80–90 a have been found, together with the shorter 22- and 11-year cycles (Suess, 1980; Sonett and Finney, 1990).

A 200-year cycle has been found in the record of δ^{10}Be concentration from the Camp Century, Greenland ice core (Beer *et al.*, 1988). These cycles are correlated with δ^{14}C tree-ring records over the last 5 ka, suggesting that the δ^{10}Be record also reflects variations in solar activity. A high correlation (0.8) has been found between δ^{14}C tree-ring records (Stuiver and Becker, 1986) and the calcium carbonate content of an Ionian Sea core over the period AD 170 to present (Castagnoli *et al.*, 1990). Spectral analysis of the ocean core also reveals a strong 200-year cycle. The calcium carbonate content of ocean sediments is controlled by biological, physical and chemical processes at the sea surface, which are themselves strongly influenced by climate. This study, therefore, provides further support for the postulated solar–climate relationship.

A 2.4 ka cycle has been identified by Hood and Jirikowic (1990) in a 7 ka composite δ^{14}C tree-ring record. This latter cycle is correlated with two independent records of solar activity: first, the auroral frequency record compiled by Schöve (1987) from Chinese and European sources; and second, the record of mean annual sunspot number since AD 1670 (Eddy, 1976). Near the maximum of the most recent 2.4 ka cycle, centred on AD 1150, the amplitude of the century-scale ^{14}C variations (the so-called Gleissberg cycles) is found to increase (Hood and Jirikowic, 1990). Thus the Maunder, Spörer and Wolf Minima are preceded by a solar activity maximum, reflecting the combined effect of the century-scale and 2.4 ka cycles.

On the basis of the studies described above, we conclude that solar forcing of the climate system may occur on both the century and the 2.4 ka time scale (Beer et al., 1988; Castagnoli et al., 1990; Hood and Jirikowic, 1990). We also note that approximate 2.4 ka periodicities have been found in oxygen isotope marine records (Pestiaux et al., 1988).

On the assumption that the δ^{14}C tree-ring record does reflect solar irradiance changes, the hypothesis that solar variability caused the observed climate fluctuations of the Holocene, including the Little Ice Age, can be investigated. Such an investigation has been carried out by Wigley (1988a) and refined by Wigley and Kelly (1990). The climate of the Holocene is represented by hemispheric and global time series of glacier fluctuations derived by combining Röthlisberger's regional time series (Figure 3.2) using approximate area weights. The dates of cold periods of glacial advance are estimated from these series. An objective filtering method is used to determine the dates of major positive ^{14}C anomalies from a reliable tree-ring record (Stuiver et al., 1986).

The episodes of cold climate and ^{14}C maxima appear to be correlated (Figure 3.5). The correlations are statistically significant at the 5% level, but a number of problems need to be addressed. Not all the δ^{14}C anomalies, for example, are associated with cold periods and vice versa. It is concluded that these results are 'highly suggestive', but not 'utterly convincing', of a δ^{14}C–climate link (Wigley and Kelly, 1990). However, the authors note the (unlikely) possibility that the δ^{14}C record reflects some process of climate change other than solar variability.

As a final step in the assessment, Wigley and Kelly use a simple energy balance model to estimate the solar forcing necessary to explain the glacial advances observed over the Holocene. By analogy with the glacial movement observed over the period of the instrumental record, it is estimated that a global cooling of 0.4–0.6°C could account for the maximum Holocene glacial advance. Depending on the assumed sensitivity of the climate to forcing, it is estimated that a solar-irradiance decrease of 0.22–0.55% over a period of about 200 a would be necessary to cause the required temperature change. This is greater than the irradiance changes predicted for the Holocene by models based on present-day relationships between sunspot activity and satellite irradiance data (Lean and Foukal, 1988). It is concluded that either the relationship between sunspots and irradiance must have varied in the past or that some additional forcing mechanism is needed to explain the Holocene glacier fluctuations (Wigley and Kelly, 1990).

These modelling results confirm the results of the most recent empirical study of ^{14}C variations and sunspot numbers (Stuiver et al., 1991). In this

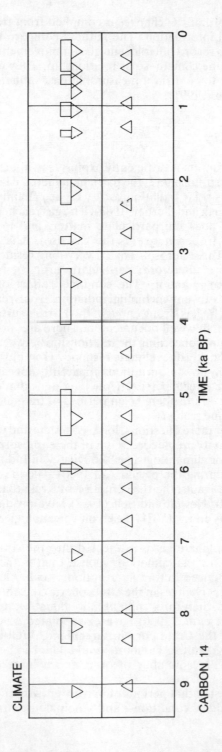

Figure 3.5 Correlations of ^{14}C anomalies and cold climate episodes. The upper half shows temperature minima (triangles) and cold periods (rectangles) identified from the record of glacier fluctuations over the last 10 ka (see Figure 3.2; and Röthlisberger, 1986). The lower half shows ^{14}C maxima (triangles). (From Wigley and Kelly, 1990.)

study, a 9.6 ka record of ^{14}C changes is compiled from tree-ring, ice-core, lake-core and coral information. The authors compare their δ^{14}C record with Röthlisberger's record of Holocene glacier movement, and also review previous studies of the climate–solar relationship. They conclude that 'we have not been able to confirm a measurable past influence of the sun on climate' (Stuiver *et al.*, 1991).

3.3.3 Volcanic activity

Where a volcanic eruption is sufficiently explosive to inject dust and sulphur dioxide gas into the stratosphere, the particulate/aerosol layer formed in the stratosphere reflects solar radiation back to space (Lamb, 1970; Rampino and Self, 1982; Deepak and Gerber, 1983). The gas reaches the stratosphere more quickly than does the particulate matter, and is less likely to be dissolved from the troposphere. As the gas rises, it is oxidized to form sulphate aerosols. These aerosols have a very long residence time and are very small. They are, therefore, particularly efficient back-scatterers of incoming short-wave radiation. The combined effect of the particulate/ aerosol layer is to prevent incoming radiation from reaching the lower atmosphere. If the effect is strong enough, the Earth's surface will cool. This mechanism is reasonably well documented. There are, however, a number of important questions concerning the relationship between volcanic activity and climate. How quickly does climate respond? How much cooling occurs? Over what area? Why do some eruptions apparently not cause cooling? How long do the climatic effects persist? How long does the 'dust veil' (Lamb, 1970) remain in the stratosphere? Can periods of frequent eruptions affect climate over long time scales?

In order to answer these questions, long, complete and well-dated records of past volcanic activity are needed. One of the earliest records of volcanic activity produced for climatologists is the Dust Veil Index (DVI) of Lamb (1970). Eruptions during the period AD 1500–1900 are included and are scaled in magnitude against the 1883 eruption of Krakatoa, which is given an index value of 1000. Newell and Self (1982) have produced a qualitative Volcanic Explosivity Index (VEI) based on volcanological data from some 8000 eruptions.

The volume of sulphate aerosols ejected into the stratosphere is more important in terms of the climatic response than is the total volume of magma or dust produced during an eruption. Acidity levels measured in polar ice cores reflect changes in the atmospheric concentration of sulphate aerosols and are, therefore, potentially valuable indicators of past eruptions. Hammer *et al.* (1980) have extrapolated a 900-year record of acidity levels from the Crete core in Greenland. Well-known eruptions, including Agung, Krakatoa, Tambora and Hekla, can be identified in this record, together with a number of other much larger but unidentified eruptions.

It has been suggested that periods of narrow growth rings in trees may be caused by the colder conditions and anomalous circulation patterns associated with volcanic eruptions (LaMarche and Hirschboeck, 1984; Baillie and Munro, 1988). The extremely narrow growth rings which may form during periods of frost damage are sometimes referred to as 'frost rings'. Absolute chronologies can be developed for tree rings, and hence they are potentially valuable records of past volcanic activity. Concentrations of very narrow growth rings over periods of less than 20 a in Irish bog oak records have been related to volcanic eruptions (Baillie and Munro, 1988). A number of the dates assigned to frost rings in bristlecone pines from the western USA have been found to coincide with eruption dates from Lamb's DVI (LaMarche and Hirschboeck, 1984). In particular, frost-ring dates are found to be within two years of Lamb's dates for the eruptions of Krakatoa (1883), Mt Pélee and La Soufrière (1902), Katmai (1912), and Agung (1963). However, comparison of frost rings and Lamb's DVI shows that not all frost events are associated with volcanic events, and vice versa. So without supporting evidence from other independent data sets, tree-ring records cannot be used as a proxy time series of volcanic eruptions.

The Santorini eruption, beginning in 1628 BC, can be identified in frost rings in trees in the western USA and may also be related to an acidity layer in the Greenland ice core record which has been given a ^{14}C date of 1645 BC. The uncertainties in the ^{14}C dates range from 50 a to 210 a and hence it is difficult to establish whether the Greenland event represents the Santorini eruption or the Avellino eruption of Vesuvius which has been dated as lying sometime in the period 1703–1617 BC (Vogel *et al.*, 1990).

Despite their dating problems, ice-core acidity records provide valuable information on changes in the frequency of volcanic eruptions over longer time scales. The Crete core from Greenland shows periods of relatively low volcanic activity, such as AD 1100–1250, and periods of higher activity, such as AD 1250–1500 and AD 1550–1700. The latter two periods lie within the Little Ice Age, while the AD 1100–1250 period can be linked to the Medieval Warm Epoch (Hammer *et al.*, 1981). It has been argued that the correspondence of these events, together with the relationship observed between glacier fluctuations and ice-core acidity records over the last 100 a, suggests that volcanic activity played an important role in the causation of the Little Ice Age (Grove, 1988).

Given the coincidence between volcanic eruptions and periods of cold climate, the remaining question is whether there is evidence that major volcanic eruptions can cause cooling of the magnitude observed during the Little Ice Age. Most of the evidence concerning the magnitude and timing of the climate response comes from superposed epoch analyses of the instrumental record (Kelly and Sear, 1984; Sear *et al.*, 1987; Bradley, 1988; Mass and Portman, 1989). A typical approach is that of Sear *et al.* (1987). The impact of five major Northern Hemisphere and four Southern Hemisphere eruptions on average hemispheric surface temperature is investigated. Month 0 indicates the time of eruption and temperatures are expressed as departures (in standard deviation units) from the mean temperature of the 12 months immediately preceding the eruption. The Northern Hemisphere eruptions are shown to have an immediate effect on temperatures in their own hemisphere, but little or no effect in the Southern Hemisphere. The Southern Hemisphere eruptions affect both hemispheres, but the greatest fall in temperature is delayed by 6–12 months. In every case, except for the Southern Hemisphere response to Northern Hemisphere eruptions, temperatures remain depressed for at least two years after the eruption. The maximum Northern Hemisphere temperature drop is 1.60 standard deviations (about 0.32°C in summer) in the second month after a (Northern Hemisphere) eruption. Thus, Sear *et al.* demonstrate that major volcanic eruptions can cause small but significant falls in hemispheric average surface temperatures over periods of two to three years.

The Sear *et al.* sample did not include the El Chichón eruption of 1982. This eruption failed to produce a clear temperature response despite evidence showing that the dust cloud reached 34–36 km into the stratosphere and that volcanic ash particulates remained at this height for at least three years (Rietmeijer, 1990). However, it seems likely that the very strong El Niño event of 1982–3 obscured the effects of El Chichón (Mass and Portman, 1989). If the El Niño effect is removed from the temperature record, it becomes easier to detect volcanic cooling (Kerr, 1989b). The five largest of nine major eruptions identified over the last 100 a have been found to cause global cooling of 0.1–0.5°C in such a corrected record (Mass and Portman, 1989). The average cooling associated with these five largest eruptions is 0.3°C, persisting for two to three years.

Most of the superposed epoch analyses have been performed using hemispheric and/or global temperature records. There is, however, some evidence to suggest that the response to volcanic forcing may show regional variation. Lough and Fritts (1987) carried out an analysis using temperature data for the period 1602–1900, reconstructed from tree-ring records from arid sites in the USA. The regional variation in the response to 24 eruptions was investigated. Central and eastern USA were found to cool (with the strongest effect in summer) while the western USA warmed (with the strongest effect in winter). It was concluded that volcanic forcing may induce changes in the atmospheric circulation (Lough and Fritts, 1987).

The interaction between volcanic and Milankovitch forcing over the last 40 ka has been investigated by Bryson (1989). A global volcanicity index was produced from radiocarbon dates of past eruptions. The geological evidence of early volcanic eruptions is often destroyed and thus the number of radiocarbon dates increases towards the present day. For this reason, a power law curve was fitted to the time series of raw counts of eruption dates. An index was then derived from this time series, by expressing the number of eruptions per century as the residuals from a power law curve. The resulting index expresses the varying frequency of volcanic eruptions. The highest frequency occurs around about 25 ka BP, while a number of smaller peaks can also be observed, for example at about 11.9 ka BP.

Bryson (1989) then combined the volcanic index with a simple model of Quaternary glaciation in which the influences of Milankovitch and volcanic forcing on irradiance are considered separately. The model is able to simulate reasonably well the observed variations in global ice volume over the late Quaternary. Events during the Late Glacial such as the Younger

Dryas are reproduced. January and July Northern Hemisphere surface temperatures are then estimated from the ice-volume time series using a heat budget model. Temperature estimates for the last 18 ka (at 3 ka intervals) are compared with the snapshot GCM modelling results of Kutzbach and Guetter (1986). The general features of the two series are considered to be broadly similar. A number of shortcomings in this analysis are identified by Bryson (1989). The limited size of the volcanicity data base, for example, means that the relative magnitude of the peaks cannot be accurately determined. Bryson's model does not include variations in the atmospheric concentration of CO_2 (see Section 2.4.2). If these were included, the agreement between Bryson's temperature estimates and the GCM results would suffer. Despite such problems, Bryson concludes that these results suggest that 'most of the century-to-millennia events are driven by variations in volcanicity'.

There is, therefore, considerable modelling and empirical evidence suggesting, but not proving conclusively, a link between climate and volcanic eruptions. It should also be borne in mind that the magnitude of the eruptions of which we have historical or contemporary knowledge is relatively small in comparison with the magnitude of eruptions in the distant past. The eruption of Tambora in 1815, for example, produced approximately 10^{11} m^3 of ash: ten times the volume produced by Krakatoa in 1883 (Kerr, 1989b). The Tambora eruption is thought to have been a major cause of the severe climatic events experienced during 1816, the 'year without a summer'. During June and July snow and frost were recorded in New England and northern Europe. In comparison, it is estimated that the eruption of Toba in Indonesia about 76 ka BP produced about 10^{12} m^3 of ash (Kerr, 1989b).

3.4 Discussion and conclusions

We concur with the statement by Grove (1988) that orbital forcing is likely to have been responsible for the 'broad sweep of environmental change throughout the Holocene, but does not account for the increasingly complex picture of climatic fluctuations which is emerging'. In the introduction to this chapter, we identified six mechanisms operating over shorter time scales (100 a to 20 ka). Three of these mechanisms have been discussed in detail: changes in ocean circulation, solar variability and volcanic activity. We conclude that the latter two mechanisms, and possibly also the first, have contributed towards the climate oscillations recorded over the last 10–15 ka. Internal mechanisms, that is air–sea–ice–land feedbacks and atmosphere–ocean feedbacks, are also likely to have contributed to episodes of glacial advance and retreat (see Section 2.4). In addition, there is evidence of changes in the atmospheric concentration of CO_2 and other greenhouse gases over these time scales. This evidence is discussed in Section 2.4.2. The effect of anthropogenic increases in the atmospheric concentration of greenhouse gases is discussed in Chapter 4. The conclusion is that a wide range of mechanisms contributes to short-term climate change.

3.4.1 Implications for future climate

This study of short-term climate change has been limited to the last 20 ka. While it is certainly of value to study the Late-glacial and Holocene period in some depth, it must be remembered that this is essentially a period of climate amelioration and warming. However, although the 'broad sweep' of change over the next 10–15 ka period may ultimately be one of cooling, it is possible that the pattern and magnitude of the short-term variability superimposed on the longer-term trends will be similar, because similar mechanisms are involved. Ideally, we would wish to identify a 10–15 ka period in the past record when climate was cooler or cooling and to examine shorter-term changes within that period. Unfortunately, such an exercise is limited by the availability of appropriate high-resolution data. A number of records exist, however, which give a general indication of the range of variation. Many of these records are discussed in detail in Chapter 7. We note their main features here.

Episodes of warmer climate within glacial periods (interstadials) can be identified in records of speleothem growth frequency from Britain (Gordon *et al.*, 1989; Section 7.2.2). Speleothem growth (or calcite deposition, here mainly in the form of stalactites and stalagmites) is related to temperature and to water availability. Hence, extensive growth indicates warm and relatively wet climate conditions typical of interglacial and interstadial conditions. Six interstadials can be identified within the Devensian glacial period, from about 80 ka to 20 ka BP. Glacial periods should not, therefore, be viewed as periods of constant severe conditions. Reconstructions of ice-sheet limits based on geological evidence reveal information about the maximum extent of ice sheets, but tell us rather less about their average extent or their variability (Porter, 1989). However, the extent of the Laurentide ice sheet is thought to have varied substantially during the last glaciation (Boulton and Clark, 1990).

Records of temperature and precipitation variability over the last 140 ka or so, reconstructed from long pollen records on the European continent, suggest that the range of variability is similar over both glacial and interglacial episodes (Guiot *et al.*, 1989; Section 7.4.1).

The relatively coarse temporal resolution of most deep-sea sediment records limits their application to studies of short-term climate variability. Some higher-resolution records are, however, available. Martinson *et al.* (1987; see Section 7.5.2), for example, have developed a high-resolution, orbitally tuned $\delta^{18}O$ chronology for the last 300 ka which has an estimated error of ±1.5 ka. This record reveals shorter-term oscillations superimposed on the major cooling and warming trends over the last two glacial–interglacial cycles.

Finally, we note the results of spectral analyses performed on oxygen isotope records from the Indian Ocean by Pestiaux *et al.* (1988; see also Section 3.1). A number of spectral peaks of less than 15 ka are observed over the last glacial–interglacial cycle (e.g., 10.3 ka, 2.7 ka and 2.5 ka). Given the relatively short length of the interglacial/interstadial episodes and the preceding warming periods, it is clear that much of the spectral power is related to variability during the episodes of glacial/stadial and cooling climate. In conclusion, we have no reason to assume that short-term variability (on time scales of 100 a to 20 ka) is less during periods of glacial or generally cooler or cooling climate.

4 Anthropogenic greenhouse gas-induced warming

4.1 Introduction

Over the last decade, both the scientific and political communities have come to recognize that human activities have the potential to cause climate change. Many scientific assessments of the anthropogenically induced greenhouse warming effect have been completed. The most recent, and perhaps the most comprehensive, is that of the International Panel on Climate Change (IPCC) (Houghton *et al.*, 1990). The highlights of the IPCC assessment were presented at the Second World Climate Conference held in Geneva in autumn 1990 (Jäger and Ferguson, 1991). The *Conference Proceedings* also contain contributions on the impacts of the enhanced greenhouse effect from world-class experts in meteorology, oceanography, agriculture, energy planning, water resource management, land use, forestry, law, health and environmental protection. These latest assessments reflect a broad consensus that enhanced greenhouse warming is a real effect (Slade, 1990).

Although orbital forcing can be recognized as the major cause of the glacial–interglacial cycles observed over the Quaternary, we consider that the major potential mechanism of climate change over the next few hundred years will be anthropogenic greenhouse gas warming.

4.2 Emissions, detection and persistence

Natural greenhouse gases in the atmosphere are necessary to sustain life: they maintain the Earth at a temperature around 33°C higher than it would be if they were not present. This is because they are largely transparent to short-wave solar radiation, but absorb long-wave radiation from the Earth. Due to human activities, such as the burning of fossil fuels, concentrations of the greenhouse gases in the atmosphere are increasing. Atmospheric CO_2 concentrations have risen from about 280 ppmv in pre-industrial times (around AD 1750) (Wigley, 1983), to about 353 ppmv today (Watson *et al.*, 1990). Concentrations of methane, nitrous oxide and the chlorofluorocarbons (CFCs) are also rising (Ramanathan *et al.*, 1985; Hansen *et al.*, 1989; Watson *et al.*, 1990). The observed changes in the atmospheric concentrations of CO_2, methane, nitrous oxide and CFC-11 are shown in Figure 4.1. The concentration of all the radiatively active trace gases in the atmosphere, taking into account their different radiative properties and residence times, is referred to as the 'equivalent CO_2 concentration'. It is estimated that CO_2 emissions account for 55% of the change in radiative forcing due to anthropogenic greenhouse gas emissions over the period 1980–1990 (Houghton *et al.*, 1990).

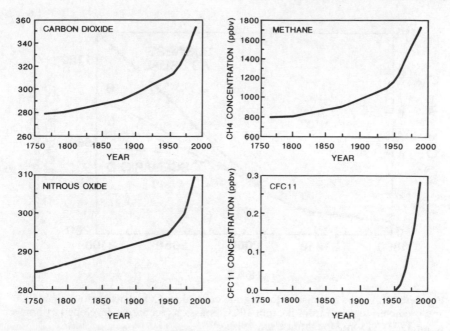

Figure 4.1 Changes in atmospheric concentrations of the radiatively active gases. (From Houghton *et al.*, 1990.)

The projection of future greenhouse gas emissions is complicated by uncertainties about their sources and sinks, and by the influence of future changes in energy, economic and environmental policies. Concern over the depletion of stratospheric ozone, for example, has led to the Montreal Protocol which limits the future use of CFCs. (It is becoming clear, however, that many of the potential replacement trace gases for the CFCs are also active greenhouse gases (Wigley, 1988b; Fisher *et al.*, 1990).) As part of the IPCC assessment, four emissions scenarios have been identified (Shine *et al.*, 1990). These are shown in Figure 4.2. The first of these is the Business-as-Usual (BaU) scenario, which assumes no policy changes. The other three scenarios are based on progressively tighter emission controls. The most stringent (scenario D), is based on a reduction of CO_2 emissions by the middle of the twenty-first century to 50% of their 1985 level. Because of the uncertainties involved, no predictions for emissions are made beyond 2100.

The majority of the model-based greenhouse warming studies have only considered the response of the climate system to a doubling of the pre-industrial equivalent CO_2 concentration, which is expected to occur sometime in the first part of the twenty-first century. One of the stated aims of this book is to assess the enhanced greenhouse effect in the context of climatic changes taking place over very long time scales. For this purpose, it is difficult, on the basis of the available evidence, to estimate the length of time over which enhanced greenhouse gas warming may persist. It could be assumed that the BaU scenario is the most realistic and that the burning of fossil

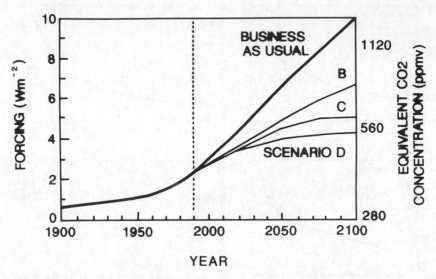

Figure 4.2 Increase in radiative forcing and equivalent CO_2 concentration since 1900 and predicted to result from the four IPCC emissions scenarios developed by Shine *et al.* (1990). (From Houghton *et al.*, 1990.)

fuels will continue until all the reserves are exhausted. Given the present political initiatives on climate change, such a scenario seems unduly pessimistic. It is, however, difficult to judge which policy options will be pursued in the future and whether any targets set will be met on time. On the other hand, it is conceivable that no effective remedial action will be taken until anthropogenic greenhouse warming has been detected with some certainty.

Although the mean globally averaged surface temperature of the Earth has increased by 0.3–0.6°C over the last 80–100 a (Jones *et al.*, 1986; Folland *et al.*, 1990), it cannot yet be claimed that enhanced greenhouse warming has been unequivocally detected (Karl *et al.*, 1989; Lindzen, 1990; Wigley and Barnett, 1990; Wigley and Raper, 1990). The observed warming lies within the natural range of variability. Statistical techniques, such as the 'fingerprint' method, are being developed in order to permit early detection (Wigley and Barnett, 1990). According to the 'fingerprint' method, a multivariate signal, with a structure unique to model-based predictions of the enhanced greenhouse effect, must be identified in the climate record. This signal might be changes in a number of different climate parameters, or changes in the same parameter, such as temperature, at a number of different places. This method satisfies two essential requirements for detection studies. First, the observed signal must be large relative to the background 'noise'. Second, the signal should be attributable to enhanced greenhouse warming alone and not to any other causal mechanism.

How long is it likely to be before we can claim that the enhanced greenhouse effect has been detected? If it is assumed that a further global warming of 0.5°C is required before detection is certain, then we will need to wait until some time between 2002 and 2047 (Wigley and Barnett, 1990). The wide range of the estimated time of detection reflects uncertainties in the emissions scenarios and in the climate models.

In Chapter 3, we discuss the influence of orbital forcing over relatively short time scales. The potential interactions between orbital and greenhouse gas-induced forcing are considered in Chapter 5. Here, we consider the ability of climatologists to provide reliable scenarios of regional greenhouse gas-induced climate change.

4.3 Regional scenario development

A stated aim of this book is to examine how regional patterns of change relate to larger-scale and long-term forcing mechanisms. Much of the work on regional climate change in relation to the enhanced greenhouse effect has focused on scenario development. Climate scenarios are not intended as predictions of future climate, but as 'internally consistent pictures of a plausible future climate' (Wigley *et al.*, 1986). They provide the basis for the evaluation of potential agricultural, ecological, hydrological, economic and social impacts of climate change. Methods have been developed for the construction of regional climate scenarios on a number of different spatial and temporal scales, and for a number of different parameters.

Three approaches to regional scenario development can be used (Wigley *et al.*, 1986; Lamb, 1987). The simplest approach is to impose an arbitrary change in certain climate variables. This approach has been used to assess the sensitivity of watersheds to changes in temperature, precipitation and evapotranspiration (Revelle and Waggoner, 1983; Flashcka *et al.*, 1987; Gleick, 1987; 1989). Typical scenarios used are ±2°C for temperature, ±10% for precipitation and ± 10% for evapotranspiration. Gleick (1987), for example, used a water-balance model to assess changes in runoff and soil moisture in the Sacramento Basin, California. With a temperature change of +2°C and a precipitation increase of 10%, the standard deviation of winter runoff increased by 22% in comparison with the control run. With the same temperature change (+2°C), but with a precipitation decrease of 10%, the standard deviation of winter runoff fell by 27%. Monteith (1981) has used a combination of mechanistic and regression models to investigate crop-yield changes in western Europe for a number of arbitrary scenarios. Cereal and potato yields were found to decrease by 5% for each 1°C temperature rise. The major disadvantage of the arbitrary approach to scenario development is that it is difficult to identify internally consistent changes for a range of climate parameters. This approach is, therefore, of limited use and is not pursued here.

The second approach involves the use of past periods of relatively warm climate as analogues of a future high-greenhouse-gas world (Webb and Wigley, 1985). This approach rests on the fundamental assumption that, given constant boundary conditions, the climate system responds in a similar way to different forcing factors. That is to say, provided the boundary conditions, such as the distribution of land and sea ice and the patterns of ocean currents, are the same, some argue that it is irrelevant whether the warming is caused by changes in greenhouse gas concentrations or, for example, in the solar irradiance: the regional climate change patterns will be the same. Potential analogues can be taken from the instrumental record or, more commonly, from the historical and geological records of palaeoclimate (Wigley *et al.*, 1986).

The Holocene thermal optimum at about 6 ka BP has been suggested as a suitable high-greenhouse-gas world analogue (Budyko *et al.*, 1987). The mean global surface temperature was then about 1°C higher than today. However, as we note in Section 3.1, the major cause of this relative warmth was the seasonal redistribution of incoming solar radiation: summer radiation increased at the expense of winter radiation. This pattern of forcing is very different from that resulting from the enhanced greenhouse effect. It is also likely that boundary conditions at 6 ka BP were substantially different from those of the present day. Modelling studies indicate that the residual Laurentide ice sheet, for example, reduced the expected warming in northwest Europe by 1–2°C (Mitchell *et al.*, 1988; Section 8.2). Similar problems are associated with the use of the previous interglacial (the Ipswichian) as a potential analogue. Even if these problems could be resolved, the lack of regional detail in the available palaeoclimate data severely limits the use of palaeoclimate analogues.

The instrumental record can, in comparison, be fairly easily used to construct consistent and very detailed regional and seasonal scenarios. This approach involves the selection of sets of warm and cold years. Differences between the two sets can then be analysed, on the assumption that they represent the changes that might be expected to occur as a result of enhanced greenhouse warming. Individual years or blocks of years can be used and it is important to choose data which maximize the temperature difference between the two sets.

The most detailed instrumental scenarios available for Europe have been constructed in the Climatic Research Unit, University of East Anglia (Lough *et al.*, 1983; Palutikof *et al.*, 1984). In selecting the data sets for analysis, the authors favoured the use of the warmest and coldest 20-year periods rather than individual years. The climatic response to greenhouse gas forcing is generally expected to evolve slowly and be associated with changes in oceanic and cryospheric conditions. The use of blocks of years should allow more accurate representation of such longer-term responses.

Spring, summer, autumn and annual temperatures were found to be up to 1°C greater over the British Isles during the warm period. The results for winter were, however, surprising. Winter temperatures fall over a large area of the European continent and southern Britain. Interannual variability was found to increase in winter during the warm period. The occurrence of colder and more variable winters is related to an increase in wintertime blocking activity, indicated by a parallel analysis of mean sea-level pressure.

The pattern of precipitation changes over the British Isles was found to be more complex. The most coherent regional pattern was seen in spring, when precipitation decreases, particularly in northwest England and Scotland. In summer, precipitation increases in northwest England and over Scotland, but is less elsewhere. In autumn, precipitation increases everywhere, but the only increases in winter are in eastern Scotland and north Wales. Elsewhere, precipitation is indicated to be lower. In general, then, a tendency was found for precipitation to decrease in all seasons other than autumn. Because of the marked spatial and temporal variability of precipitation, however, these precipitation scenarios are not considered to be reliable.

The Climatic Research Unit instrumental scenarios are based on two sets of years which differ in their mean temperatures by only about 0.4°C. This contrast is very small in comparison with the model projections for temperature increases associated with a doubling of CO_2. Therefore, such instrumental scenarios can only be used as indicators of climatic conditions during the early phase of a greenhouse gas-induced warming. Instrumental scenarios are based on the assumption that no anthropogenic greenhouse warming has yet occurred. In the example of the Climatic Research Unit scenarios, however, the block of warm years is more recent than the block of cold years. Hence these scenarios may in fact represent a greenhouse gas-induced climate change that has already occurred, in which case the scenarios cannot be used for predictive purposes unless we assume that the changes will be linear.

One advantage of the instrumental scenario approach is that a variety of instrumental data may be used. This approach has, for example, been employed to develop scenarios of cloud amount in a warming world (Henderson-Sellers, 1986; Henderson-Sellers and McGuffic, 1989). The differences between cloud cover observed in 1901–1920 (the cold period) and in 1934–1953 (the warm period) were calculated. Annual and seasonal scenarios for Europe (Henderson-Sellers, 1986) and for sub-continental India and Canada (Henderson-Sellers and McGuffie, 1989) were then produced. These scenarios are discussed further in Section 4.4.3.

The third method of scenario construction, and the most widely used, is based on output from climate models and is discussed in detail in the next section.

4.4 The reliability of general circulation models

4.4.1 Introduction

The most complex climate models and those most widely used in studies of climate change are the three-dimensional GCMs (MacCracken and Luther, 1985; Schlesinger and Mitchell, 1987; Houghton et al., 1990). GCMs and numerical weather prediction models grew from similar origins in the 1950s. The aim of GCMs is to simulate the full three-dimensional character of the climate, making them the most sophisticated existing models of the atmosphere. They solve the primitive equations that describe the movement of energy and momentum, and the conservation of mass and water vapour. Physical processes, such as cloud formation, and heat and moisture transport within the atmosphere, and between the atmosphere and the surface, are also described. Atmospheric conditions are specified at a number of 'grid points' on a regular grid over the Earth's surface, and at several levels in the atmosphere. The primitive equations are then solved at each grid point, using numerical techniques. Although various methods are available, all models use a time-step approach and an interpolation scheme between grid points (Henderson-Sellers and Robinson, 1986; Washington and Parkinson, 1986).

The current generation of GCMs model the equilibrium rather than the transient response of climate to greenhouse gas forcing. For the control run, the atmospheric concentrations of CO_2 are set at pre-industrial levels and the model is then run until it reaches equilibrium. This procedure is repeated using doubled CO_2 concentrations, the perturbed run. The differences in the results for the perturbed and control runs are taken to be the climate change due to the enhanced greenhouse effect. In reality, atmospheric concentrations of the greenhouse gases are steadily increasing with time, and the atmospheric response lags behind, due to the oceans' thermal inertia. As discussed in Section 4.4.6, under this circumstance of transient response, the patterns of climate change may be very different from those predicted by GCMs for the equilibrium response.

Much of the published literature relates to five individual GCMs, each of which is known by the name of the group responsible for its development: the UK Meteorological Office (UKMO); Goddard Institute of Space Studies (GISS); National Center for Atmospheric Research (NCAR); Geophysical Fluid Dynamics Laboratory (GFDL); and Oregon State University (OSU). The NCAR model is also known as the Community Climate Model (CCM).

In order to demonstrate some of the problems which arise when trying to use GCM grid-point data in climate impact studies, we have obtained temperature and precipitation data for western Europe from each of these models. Seasonal output from control and CO_2-perturbed model runs has been obtained and is evaluated in Section 9.2. The key characteristics of the models developed by each of the five modelling groups are outlined in Table 4.1. Details are given of the model runs for which we have data, not necessarily the most recent model run. It should be stressed that there are many different versions of each model.

All of these models are described as having 'realistic' land/ocean distribution and topography. Sea ice and snow are calculated from parameters generated within the model, rather than being prescribed from observational data sets. Clouds are calculated in each atmospheric layer in all models. With a doubling of CO_2, the models predict an increase in global mean surface temperature in the range 2.8–5.2°C and an increase in global precipitation of the order of 7–15%. This temperature range is broadly consistent with the best estimates (1.5–4.5°C) of a global temperature rise associated with a doubling of CO_2 identified by reviews of all equilibrium GCMs available in the mid-1980s. (MacCracken and Luther, 1985; Bolin *et al.*, 1986).

Improvements in computing speed and power over recent years have

Table 4.1 Key characteristics of the five general circulation models (GCMs) discussed in the text. Output from these models for northwestern Europe is evaluated in Section 9.2.

	UKMO	GISS	NCAR	GFDL	OSU
LAT×LONG	5° × 7.5°	7.83° × 10°	4.5° × 7.5°	4.5° × 7.5°	4° × 5°
VERTICAL LAYERS	11	9	9	9	2
INSOLATION	annual and diurnal	annual and diurnal	annual	annual	annual
OCEAN MODEL	prescribed heat exchange	prescribed heat exchange, varying mixed layer	mixed layer	mixed layer	6-layer ocean GCM
GLOBAL WARMING 2×CO_2	5.2°C	4.2°C	3.5°C	4.0°C	2.8°C
GLOBAL PRECIPITATION CHANGE 2×CO_2	+15%	+11%	+7.1%	+8.7%	+7.8%
FOR MODEL DESCRIPTION, SEE	Wilson and Mitchell, 1987	Hansen et al., 1984	Washington and Meehl, 1984	Wetherald and Manabe, 1986	Schlesinger and Zhao, 1989

allowed all groups to implement substantial improvements to their models. These improvements include higher resolution and the introduction of ocean heat transport, a diurnal cycle, cloud water and variable cloud radiative properties. Groups based in Australia, Canada, China, France, Germany, Japan and the former USSR are all developing GCMs (Cubasch and Cess, 1990; Cess et al., 1991). Many of these new GCMs are at the developmental stage and have not yet been used to investigate the impacts on climate of changes in the atmospheric concentration of greenhouse gases. The key characteristics of many of the most recent versions which have been used to study the enhanced greenhouse effect are summarized in table form by Cubasch and Cess (1990, Table 3.2). Despite the great advances which have been made in climate modelling, a number of major shortcomings remain. These are summarized in Table 4.2 and are discussed in Sections 4.4.2–4.4.6.

4.4.2 Spatial resolution

In comparison with the requirements of climate impact assessment and with the scale of many climate processes, all models suffer from poor horizontal resolution (Gates, 1985; Lamb, 1987; Goodess and Palutikof, 1990; Giorgi

Table 4.2 Outstanding problems with GCMs run in equilibrium mode.

- low spatial resolution
- poor representation of sub-grid-scale processes
- lack of relief and unrealistic geography
- poor representation of feedback processes particularly cloud feedbacks
- failure to reproduce details of present-day climate
- regional discrepancies between the predicted $2 \times CO_2 - 1 \times CO_2$ changes, particularly for parameters such as precipitation and soil moisture
- do not take into account time-dependent effects, such as thermal inertia of the oceans or transient nature of greenhouse gas forcing

---

and Mearns, 1991; Grotch and MacCracken, 1991). The typical size of a single grid square (see Table 4.1) ranges from 8° latitude by 10° longitude (GISS GCM, Hansen *et al.*, 1984) to 4° latitude by 5° longitude (OSU GCM, Schlesinger and Zhao, 1989). The higher horizontal resolution of the latter model is only achieved at the expense of a simple layer structure for the atmosphere (two rather than nine atmospheric layers). This particular model, however, also includes a multi-layer ocean. In recent years, a limited number of high-resolution models have been run. Output from three such models (versions of the Canadian Climate Center, GFDL and UKMO GCMs) is presented in the IPCC report (Cubasch and Cess, 1990; Mitchell *et al.*, 1990). (Output from the perturbed runs of these three models, for southern Europe, is discussed in Section 9.2.2). The UKMO model, for example, has been run with a horizontal resolution of 2.5° latitude by 3.75° longitude and a vertical resolution of 11 layers (Mitchell *et al.*, 1989). Thus, the horizontal resolution of GCMs is of the order of 300–1000 km.

The representation of climate parameters as simple grid-point averages is unrealistic. In most GCMs, for example, the total precipitation estimated at each grid point is spread evenly over the grid square. Thus precipitation tends to occur as 'drizzle', rather than as convective local storm events. In reality, the proportion of the rainfall lost as runoff depends in part on the intensity. GCMs, therefore are likely to underestimate local runoff and overestimate soil moisture storage and evaporation. The greatest effects are expected in tropical forest areas and have been demonstrated using the Biosphere–Atmosphere Transfer Scheme (Pitman *et al.*, 1990). When precipitation is more realistically distributed in space and time by the model, the climate regime changes from one dominated by evaporation to one dominated by runoff.

Low spatial resolution affects the ability of the model to simulate sub-grid-scale processes such as convective rainfall. It also makes it difficult to develop meaningful regional scenarios of climate change due to the enhanced greenhouse effect. To overcome the latter problem, techniques for the estimation of sub-grid-scale values have been developed. The need to interpolate sub-grid-scale values from coarse resolution grids (or point values from area means) is often referred to as the 'climate inversion' problem (Gates, 1985), and has been tackled by two groups of scientists using present-day climate data. The first statistical techniques to be proposed were developed by Kim *et al.* (1984), using climate data from Oregon in the USA. The method is based on principal component and regression analysis of seasonal temperature and precipitation observations.

Principal component regression equations are constructed in which observed station temperature and precipitation variations are explained by regionally averaged variations in a range of climate variables which may include temperature, precipitation and atmospheric pressure. Then, the regionally averaged (predictor) variables are replaced by GCM $2\times CO_2$–$1\times CO_2$ differences, averaged across a number of grid points. Finally, the predicted station values can be used to produce a contoured map showing sub-grid-scale climate perturbations.

The initial Kim *et al.* analysis only utilized spatial variations in the amplitude of seasonal cycles of temperature and precipitation in the regression equations and is, furthermore, considered inadequate for the production of physically reasonable results for precipitation (Wilks, 1989; Wigley *et al.*, 1990).

Wilks (1989) has extended the method by the use of daily maximum and minimum temperature and precipitation data, by the introduction of a stochastic precipitation model and by the treatment of data intercorrelations. In this study, data are taken from three regions of Kansas, Iowa and Dakota, equivalent to three grid squares of the NCAR GCM. The improved method is based on rotated principal component analysis of the correlation matrix of observed station data with area mean data. The stochastic model allows the observed locations and intensities of convective precipitation (a sub-grid-scale process) to be simulated as random phenomena. This method does, however, require a dense network of precipitation data to match the spatial scale of convective precipitation. Wilks considers it unwise to interpolate to points not used in developing the correlation matrix.

A slightly different approach has been adopted by Wigley *et al.* (1990). This particular method is also developed using data from Oregon, but interannual climate variations rather than seasonal cycles of the parameters are employed. In addition, zonal and meridional pressure gradients are included as predictors in the regression equations.

While both techniques (Wilks, 1989; Wigley *et al.*, 1990) produce useful results, a question mark concerns their transferability. These techniques have been developed for regions of the USA where the local climate pattern is determined by topographic/thermal forcing factors which are themselves relatively simple in form. In Oregon, for example, the coastline and the principal mountain ranges run north–south. The effects of the same forcing factors in other regions, such as the British Isles, may be much more complex. In particular, meteorological data from a very dense network of stations may be required in order to capture the spatial complexity of both the forcing factors and the climate itself.

The studies described above indicate that, in some parts of the world at least, it is possible to establish meaningful relationships between present-day area mean values and local station values of climate variables. In order to use these techniques in studies of climate change it is necessary to assume that these relationships will be unchanged. This is not an easy assumption to test. Wigley *et al.* (1990) do, however, put forward two arguments in support. First, they say, the estimated mean values of climate variables such as temperature and precipitation in a high-greenhouse-gas world lie within the range of present-day extremes. Therefore, the statistical relationships developed on the basis of the present-day data can be applied to the investigation of greenhouse gas-induced changes without extrapolating beyond the bounds of the data used in the construction of the regression model. Second, they argue that the relationships must reflect the influences of orography, geography and land-surface characteristics and that these are unlikely to change. It is certainly reasonable to assume that no major orographic or geographical changes will occur over the time scale of greenhouse warming. Some minor changes in geography may occur due to coastal flooding as sea level rises. It is, however, less reasonable to assume that changes in vegetation and agriculture, affecting land-surface characteristics, will not occur.

Whether or not valid and useful statistical relationships can be developed for constructing sub-grid-scale climate scenarios over a region such as the British Isles, the physical representation in GCMs of sub-grid-scale processes, such as convective precipitation and cloud formation, remains poor. Inevitably this must affect the accuracy of the predicted grid-point values.

Rind (1988) has demonstrated how model resolution affects two key processes in control runs of the GISS GCM: moist convection and the transfer of kinetic energy. In the fine grid model (4° latitude by 5° longitude) more penetrative convection but less overall convection occurs than in the coarse grid model (8° latitude by 10° longitude). This leads to differences in the wind and temperature structures of the two model control climates. The fine mesh model simulates stronger winds, more evaporation and a stronger hydrological cycle than the coarse mesh model. Synoptic features, such as eddy momentum transport and the Ferrel cell, are considered to be more realistic in the fine mesh control run. The coarse mesh model has more realistic rainfall rates in the tropical regions, probably due to the fact that the moist convection scheme used in both models was developed for this resolution. Rind concludes that 'overall, then, the finer resolution produces mixed benefits in the present climate simulation, and if the criteria [*sic*] for correct prediction of climate change is a more accurate control run, it is not clear which model is to be preferred'.

Comparisons between the fine and coarse mesh models have also been made for $2 \times CO_2$ and 'ice age' perturbed runs (Rind, 1988). The fine grid $2 \times CO_2$ run shows a greater decrease in high-level cloud cover and greater increases in temperature, surface winds, precipitation and penetrative convection. Strong differences in regional climates arise because of the change in model resolution. In the $2 \times CO_2$ experiments, for example, soil moisture changes over the Sahel are negative in the fine mesh model and positive in the coarse mesh model. In the 'ice age' experiments the estimated decrease in temperature over the eastern USA and southern Canada is over 3°C greater for the fine mesh model than for the coarse mesh model.

However, over Africa and Asia, the coarse mesh model estimates temperature decreases about 1°C greater than the fine mesh model (Rind, 1988).

Climatologists describe the geography of the present generation of GCMs as 'realistic' but local detail is often lacking. The Mediterranean and Hudson Bay, for example, are generally treated as closed lakes. The Panama Isthmus, together with islands such as Japan and New Zealand, are missing from some models. Orography is also highly simplified, restricted to major features such as the Alps, the Tibetan Plateau and the Rockies. In Section 9.2, we demonstrate that direct comparisons between the simplified GCM geography and real-world geography cannot be made.

A potentially promising approach to the spatial resolution problem involves the use of a limited area model (LAM) nested within a GCM (Giorgi and Mearns, 1991). This approach has been demonstrated over Europe using coarse-resolution GCM data (7.5° × 4.5°) to drive a LAM with a grid-point spacing of 70 km (Giorgi et al., 1990). Given the very large computing requirement of this model, and the early stage of development, it has initially been run for a 10-year period and for January conditions only. The preliminary results are encouraging, but the development of regional scenarios for the western USA, Europe, the Great Lakes region and southeast Australia remains a long-term goal.

4.4.3 Model sensitivity and feedback effects

The five GCMs described in Table 4.1 predict a rise in global temperature of anywhere between 2.8°C and 5.2°C with a doubling of CO_2. The magnitude of the predicted CO_2-induced warming can be taken as an indication of the model's sensitivity. The UKMO GCM, for example, may be considered the most sensitive because it produces the largest estimate of CO_2-induced warming (5.2°C), whereas the least sensitive is the OSU GCM, with a global temperature change of 2.8°C. A number of recent studies have attempted to identify those attributes of the models which lead to such divergent results. The model differences may arise for three main reasons (Cess and Potter, 1988): differences in control climates; differences in the initial direct radiative forcing; and physical differences in climate feedback processes.

Comparative studies indicate that a model's response to enhanced CO_2 levels is highly dependent on the global mean control climate. The colder the control climate is, the greater the model sensitivity (Spelman and Manabe, 1984; Cess and Potter, 1988). This relationship is demonstrated graphically in Figure 4.3. It has been concluded that 'perhaps most of the differences in CO_2-induced global warming, as predicted by five contemporary GCMs, are due to differences in their control climate' (Cess and Potter, 1988). Ideally, models should be developed to minimize the discrepancies between control climates and to separate out and quantify differences due to radiative forcing and to feedback effects.

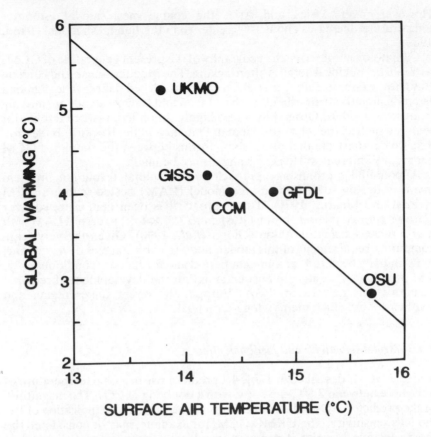

Figure 4.3 Relationship between GCM control climate temperature and predicted global warming. Global warming due to $2 \times CO_2$ is plotted against global mean $1 \times CO_2$ surface air temperature for five GCM simulations. (From Cess and Potter, 1988.)

Cloud feedback effects

In a follow-up study, a comparison has been made of 19 GCM experiments in which sea-surface temperature (SST) changes of $\pm 2°C$ were used as a surrogate for enhanced greenhouse gas forcing (Cess *et al.*, 1989; 1990). Models were run using identical boundary conditions. A threefold variation in the sensitivity of model climates to higher SSTs was found. Most of this variability was attributed to differences in cloud–climate feedback processes. These differences may also account for part of the observed differences in control climate. The effects of clouds were isolated by separately averaging the clear-sky (i.e. no cloud cover) and overcast-sky (i.e. with cloud cover) top of the atmosphere fluxes. Cloud cover decreased with warming in each model, but the associated cloud feedback varied from a modest negative effect to a strong positive effect (Figure 4.4). Although

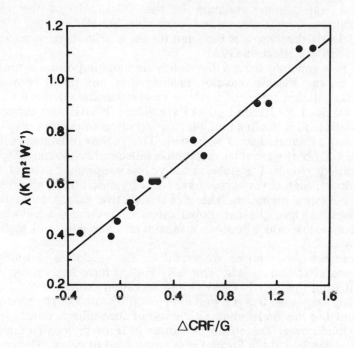

Figure 4.4 Relationship between cloud feedback effects and sensitivity to sea-surface temperature changes. The global sensitivity parameter (λ) is plotted against the cloud feedback parameter $\Delta CRF/G$ for 14 GCM simulations. The solid line represents a best-fit linear regression.

$$\lambda = \frac{1}{\Delta F/\Delta T_s - \Delta Q/\Delta T_s}$$

$$CRF = F_c - F + Q - Q_c,$$

$$G = \Delta F - \Delta Q$$

where F = global mean emitted infrared terrestrial fluxes at the top of the atmosphere
  Q = net downward solar fluxes at the top of the atmosphere
  F_c = as F, but for clear-sky fluxes only
  Q_c = as Q, but for clear-sky fluxes only
  T_s = global mean surface temperature
  G = direct radiative forcing

(From Cess *et al.*, 1990.)

the net cloud feedbacks were found to be similar in a number of models, this masked considerable variation in the magnitude of the individual contributions from stratiform and convective clouds (Cess *et al.*, 1990). This study clearly demonstrates the need for more reliable parametrization of cloud feedback effects in GCMs.

It is now generally agreed that clouds are more important in terms of the effect on the Earth's radiation balance than any direct aerosol effect. Modelling studies are benefiting from newly available satellite data. Results from the Earth Radiation Budget Experiment (ERBE), for example, have demonstrated for the first time that the net effect of clouds today is to cool the Earth (Ramanathan *et al.*, 1989). This is shown diagrammatically in Figure 4.5. Short-wave solar radiation is reflected back to space by low- and mid-altitude clouds. The global radiative loss is estimated to be 44.5 W m^{-2}. In contrast, high cirrus clouds have a greenhouse effect, absorbing long-wave terrestrial radiation. The global radiative gain is estimated to be 31.3 W m^{-2}. Thus, the net global effect of present-day cloud cover is a radiative loss which reaches a maximum over mid- and high-latitude oceans.

These new observations are useful for the validation of water vapour feedback effects in GCMs. The ERBE data have been compared with output from the NCAR CCM (Kiehl and Ramanathan, 1990). The total outgoing long-wave flux is found to be overestimated by the model. This is attributed to the model underestimation of atmospheric water vapour and high cloud cover. The effects of clouds in three regions (Indonesia, the North Atlantic and the Pacific), are considered in detail. The Indonesian data show that short-wave and long-wave radiation effects cancel each other out. The model, however, predicts net cooling due to the greater reflection of short-wave radiation back to space. Over the Atlantic and Pacific, the model underestimates the expected cooling by a factor of 2.

Observations such as those collected during the ERBE have implications for the study of cloud feedbacks and their relationship to past and future climates. It has been suggested, for example, that the southward shift of mid-latitude storm tracks (and their associated low- and mid-altitude clouds) during glacial periods may have amplified the cooling and so increased continental glaciation (Ramanathan *et al.*, 1989). The direction of cloud feedback effects depends critically on factors such as cloud area, altitude, the proportion of different cloud types, and water and ice content (Rind, 1988; Schlesinger, 1988). Work is now going forward to improve the modelling of cloud processes, including convection and evaporation (Cess and Potter, 1988; Charlock *et al.*, 1988; Schlesinger and Oh, 1988; Slingo and Slingo, 1988).

Analysis of long-term records suggests that cloud amounts may increase in a warming world, at least over some land areas (Henderson-Sellers, 1986; Henderson-Sellers and McGuffie, 1989). In a scenario study, cloud amount data from two 20-year periods were compared. Total cloud amount was found to increase over practically all North America, most of the Indian subcontinent and parts of Europe in the warm period (Henderson-Sellers and McGuffie, 1989). Observation-based studies, such as this, may prove

Long wave: $+31.3\,Wm^{-2}$ Short wave: $-44.5\,Wm^{-2}$

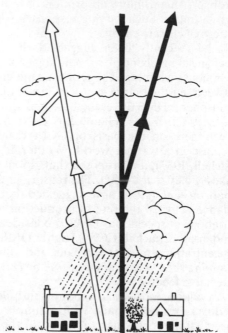

POSITIVE FEEDBACK
Absorption of long-wave radiation by high-level cirrus cloud.
Height of cloud top is important.

NEGATIVE FEEDBACK
Reflection of short-wave radiation by low-/mid-altitude cloud. Maximized over mid-/high-latitude oceans.
Water/ice content of clouds is important.

Figure 4.5 Diagrammatic representation of the effects of clouds on climate. (Data from Ramanathan *et al.*, 1989.)

useful in the evaluation of GCM results, although they have their own inherent problems. Long-term observations of cloud amount are only available over land areas and not over the oceans, so that it is difficult to assess global impacts. In addition, the observations are of *total* cloud cover only and do not distinguish between cirrus and lower-altitude cloud cover. In order to estimate the overall direction of cloud feedback effects, it is essential to know the proportions of these two cloud types and to know something of their detailed characteristics such as droplet size.

The relative importance of the two cloud feedback effects may change in a high-greenhouse-gas world. Furthermore, the geographical distributions of areas of net negative and net positive cloud feedback effects is likely to alter. Results from an experiment carried out with the UKMO GCM, for example, suggest that negative cloud feedbacks associated with increased low-level cloud will dominate at mid-latitudes (Mitchell *et al.*, 1989). Satellite observations show that, at the present time, the net effect of cloud cover at mid-latitudes is a radiative loss (Ramanathan *et al.*, 1989; Kiehl and Ramanathan, 1990). The results of Mitchell *et al.*, therefore, imply that the direction of the cloud feedback effects will not change (although the intensity of the effect on the radiation balance may). The model-based study of Wetherald and Manabe (1986) indicates a decrease in low-level cloud, producing a positive feedback (due to less reflection of short-wave radiation), and an increase in high-level cloud, also producing a positive feedback (due to more absorption of long-wave radiation), particularly at high latitudes. Such varying responses will affect both the magnitude and pattern of climate change.

It has recently been suggested that a negative biological feedback mechanism involving planktonic algae and cloud albedo could help to offset greenhouse warming. The proposed mechanism is based on the relationship between seasonal surface insolation and dimethylsulphide (DMS) flux to the atmosphere: DMS fluxes increase in response to increased insolation (Bates et al., 1987). If this relationship is extended to the enhanced greenhouse effect then, on the basis of GCM results showing that cloud amounts decrease in a warmer world (Wetherald and Manabe, 1986; Wilson and Mitchell, 1987), it is proposed that global warming will lead to higher DMS fluxes (Foley et al., 1991). The reason for the increased flux is uncertain, but Charlson et al. (1987) have suggested that it could be due to changes in the balance between the rate of production and removal in the ocean, and to variations in speciation. DMS is considered to be a major source of cloud condensation nucleii (CCN). Higher DMS fluxes will lead to increased CCN concentrations over the oceans and thus a higher cloud albedo. More incoming short-wave radiation will be reflected back to space, thus creating a negative feedback.

An initial estimate of the magnitude of the proposed DMS–cloud feedback has been made with a simple model incorporating empirically based parametrizations to account for the biological control of CCN (Foley et al., 1991). The direct effects of this negative feedback on short-wave radiation reduce the imposed changes in solar forcing in the model by less than 7%. The operation of other feedback effects may also be affected on a similar scale. It is concluded that these preliminary results indicate that the proposed DMS–cloud feedback effect will not have a significant influence on the response to enhanced greenhouse warming (Foley et al., 1991). The proposed DMS–cloud feedback link remains controversial (Schwartz, 1988; Savoie and Prospero, 1989; Wigley, 1989; Ghan et al., 1990; Monaghan and Holdsworth, 1990); further observational and modelling studies are needed.

Thus, cloud feedbacks remain one of the greatest uncertainties in climate modelling. They are not, however, the only feedbacks which are likely to affect the response of climate to enhanced greenhouse gas forcing. Potential feedbacks are summarized in Table 4.3.

Water vapour feedbacks
One of the most important feedback mechanisms involves water vapour. Warming leads to enhanced evaporation and higher concentrations of atmospheric water vapour which, because water vapour is itself a greenhouse gas, in turn leads to further warming. This mechanism has been widely regarded as a relatively simple positive feedback which can readily be parametrized in climate models (Cess, 1989; Mitchell et al., 1990). It is the water vapour feedback, rather than the direct effects of increased greenhouse gas concentrations, which is responsible for much of the warming observed in the perturbed runs of GCMs (Hansen et al., 1984).

Table 4.3 Climate feedback effects and the responses to enhanced greenhouse warming.

Feedback	Direction	Reference
Clouds (increase in high-level clouds)	positive	Platt, 1989; Ghan *et al.*, 1990
Clouds (increase in low-level clouds)	negative	Mitchell *et al.*, 1989; Slingo, 1990
Clouds (DMS and CCN)	negative	Charlson *et al.*, 1987
Clouds (net effect of changes)	positive or negative	Cess *et al.*, 1989
Water vapour	positive	Cess, 1989; 1991; Del Genio *et al.*, 1991
Snow cover	weak negative to strong positive	Cess *et al.*, 1991
Sea ice–surface albedo	positive	Dickinson *et al.*, 1987; Ingram *et al.*, 1989; Covey *et al.*, 1991
Biogeochemical ocean CO_2 uptake and release:	positive	Lashof, 1989
terrestrial biota: (albedo, carbon storage, CO_2 fertilization)	positive	
CH_4 emissions from wetlands and paddy fields: tropospheric chemistry:	positive negative	

It has recently been suggested that sinking air in regions surrounded by deep cumulus clouds may result in the local drying of the upper troposphere, thereby eliminating or reversing the water vapour feedback in some regions (Lindzen, 1990). This view is not supported by recent modelling and observational evidence (Cess, 1991; Del Genio *et al.*, 1991; Rind *et al.*, 1991), which confirms the water vapour feedback as the largest single positive feedback effect.

Rind *et al.* (1991) have analysed water vapour data for January 1985 to July 1989 collected by satellite as part of the ERBE. First, they compared summer and winter observations in the middle and upper troposphere. This comparison shows that, as the hemisphere warms, so the amount of convection and water vapour above 500 mb increases. Second, they compared data from the west Pacific, which is dominated by convection, with data from the east Pacific, which is a largely non-convective region. In comparison with the east Pacific, the middle and upper troposphere is found to be moister in the west Pacific. These observations confirm that increased convection is associated with increased water vapour content in the upper troposphere and, therefore, with a stronger positive feedback effect.

The satellite data have also been used to validate the representation of water vapour feedbacks in GCMs (Del Genio *et al.*, 1991; Rind *et al.*, 1991). Del Genio *et al.* (1991) have used two versions of the GISS GCM to investigate water vapour changes using SST changes of $\pm 2°C$ as a surrogate for enhanced greenhouse gas forcing (Cess *et al.*, 1989; 1990). The two model versions differ only in their representation of cumulus and stratiform clouds. In both models, the upward transport of moisture is enhanced and the accumulation of water vapour and ice at the cumulus cloud tops is increased with warming. These observations are consistent with the satellite observations (Rind *et al.*, 1991). The global warming due to the water vapour feedback is estimated to be 2.69°C and 2.98°C, respectively, in the two models (Del Genio *et al.*, 1991). The similarity of these results suggests that temperature, and not the fine tuning of cloud parametrization in the second model, is the main control on the magnitude of the feedback.

Albedo feedbacks
The behaviour of snow–climate feedbacks (Table 4.3) in 17 GCMs, developed by groups based in nine different countries, has recently been investigated (Cess *et al.*, 1991) using a similar methodology to that employed in the cloud-feedback study discussed above (Cess *et al.*, 1989; 1990). An SST change of ±2°C was used as a surrogate for enhanced greenhouse gas forcing and the simulation was carried out for perpetual April conditions. This study shows that the conventional view that snow cover will be reduced in a warmer world and that, therefore, more solar radiation will be absorbed due to the change in albedo, is simplistic (Cess *et al.*, 1991). The net effect of the snow feedbacks is found to differ markedly from model to model, from a weak negative to a strong positive effect.

The occurrence of a weak negative feedback in five of the GCMs is attributed to interactions with clouds (Cess *et al.*, 1991). In these cases, clouds are thought to counteract the positive snow feedback effect. For example, an increase in cloud cover over a region of bare soil would counteract the reduction in albedo caused by snow retreat. An indirect effect on long-wave radiation is also seen in some of the models. Where snow retreat causes a steeper atmospheric lapse rate, less long-wave radiation is emitted from the colder atmosphere. This study demonstrates that, as in the case of cloud feedbacks, net snow feedbacks reflect a complex balance of direct and indirect radiative effects (Cess *et al.*, 1991).

As warming occurs, so sea ice melts, surface albedo falls and more radiation is absorbed by the Earth's surface, thus intensifying the initial warming. This is the sea ice–surface albedo feedback effect. The strength and realism of this feedback in GCM simulations has also been assessed recently (Ingram *et al.*, 1989). Differences arising from changes in the parametrization of this feedback in the UKMO GCM are evaluated. The magnitude of the feedback in a range of models has also been assessed. The feedback effect is positive in all models: the 'standard' UKMO model (Wilson and Mitchell, 1987) is found to have the lowest albedo effect and the GFDL model the largest.

An upper limit for the contribution of the sea ice–albedo feedback to enhanced greenhouse warming has been estimated (Covey *et al.*, 1991). In this experiment, performed with the NCAR CCM, the increased absorption of solar energy at the Earth's surface, if all sea ice melts, is calculated. In order to isolate the climatic effects of the albedo change, parameters such as temperature and cloud cover are left unchanged over each model run (or integration). Integrations are carried out at several time points over the seasonal cycle. The global annual radiation increase is estimated to be 2–3 $W m^{-2}$ (Covey *et al.*, 1991). Because the model tends to underestimate cloud amount, the lower end of this range, 2 $W m^{-2}$, is taken to be a realistic real-world upper bound. In comparison with the direct radiative forcing effects of $2 \times CO_2$, this increase is significant. For the reduction in sea ice likely to be associated with $2 \times CO_2$ conditions, however, the sea ice–albedo effect can be expected to be considerably less than the water vapour feedback (Covey *et al.*, 1991).

Biosphere feedbacks

In addition to the cloud, water vapour, snow and sea-ice feedbacks discussed above, there are a number of feedbacks which involve interactions with the biosphere (Table 4.3). Changes in both flora and fauna can affect the rate of emission of greenhouse gases and their uptake. It has, for example, been suggested that methane emission rates from wetlands and rice paddy fields will increase due to enhanced greenhouse warming (Lashof, 1989). Methane emissions from these regions may, however, be exceeded by those from melting permafrost in higher latitudes (Watson *et al.*, 1990).

The role of the oceans in the global carbon cycle is currently the focus of intensive research. Emerging modelling results suggest that the oceans' biological carbon pump may play an important role in the response to climate change (Heinze and Maier-Reimer, 1989).

Recent modelling work also demonstrates the importance of land-surface parametrizations in GCMs (Lean and Warrilow, 1989; Warrilow and Buckley, 1989). Variations in surface albedo and roughness, for example, influence the energy and moisture exchanges between the surface and the atmosphere (Warrilow and Buckley, 1989). The importance of these interactions is demonstrated by a GCM simulation of the regional climatic impact of Amazon deforestation (Lean and Warrilow, 1989).

A high-resolution version of the UKMO GCM (2.5° latitude by 3.75° longitude) was used to explore the effect of replacing all Amazon tropical forests with savannah and pasture (Lean and Warrilow, 1989). The local climate response to deforestation is dominated by a weakened hydrological cycle, with less precipitation and evaporation, and by an increase in surface temperature. The reduction in evaporation is largely caused by a lower surface roughness once the tree cover is removed. Although lower evaporation rates contribute to the decrease in precipitation, the principal cause is related to a rise in the albedo. This reduces the amount of energy available for the upward turbulent transfer of latent energy, and hence the vertical moisture flux is further reduced. Together, the changes in albedo and roughness are responsible for a 20% decrease in precipitation. These results are qualitatively similar to those of other studies (Dickinson and Henderson-Sellers, 1988; Mitchell *et al.*, 1990; Shukla *et al.*, 1990). The consensus view is that the direct effects of tropical deforestation on regional climate may be large, but that the impact on global climate will be relatively small. In a recent expert assessment it is argued that removal of all the tropical forests could only warm the global climate by about 0.3°C (Mitchell *et al.*, 1990).

The interactions between ecological and climate processes are complex. Many of the crucial processes are still poorly parametrized in GCMs. The typical soil moisture scheme, for example, consists of a 15 mm 'bucket' which fills or empties according to a highly idealized relationship between precipitation, evaporation and runoff (Warrilow and Buckley, 1989). Improved parametrizations of land-surface hydrology which incorporate realistic physical mechanisms and sub-grid-scale variability are, however, being developed (Entekhabi and Eagleson, 1989; Warrilow and Buckley, 1989).

4.4.4 Model validation

In order to evaluate output from the perturbed runs of GCMs, output from the control runs must be validated against independent observations of present-day climate. The main purpose of validation is to identify systematic errors, particularly those which may be common to a number of models (Gates *et al.*, 1990). Any identified errors or biases must be taken into account when using GCMs to investigate future climate conditions.

One of the principal methods of model validation is to compare control run results with observational data sets. Because of the coarse resolution of GCMs, and because the output from the model runs is generated at each grid point only, it is not always appropriate to compare the results directly with station observations of climate variables. In addition, station observations are affected by local factors such as altitude (Jones *et al.*, 1985). Gridded data sets have been produced from observations for a limited number of variables, for example the combined land and sea temperature data set produced by Jones and colleagues (see Folland *et al.*, 1984; Jones *et al.*, 1985; 1986). It is not, however, possible to validate GCM temperature simulations, since the control runs use temperature as a boundary condition. Gridded data sets of precipitation and pressure are, therefore, more appropriate for validation purposes. The preparation of gridded data sets for parameters such as precipitation and pressure is constrained by the lack of comprehensive observations over the oceans. However, in some cases it may be possible to compare GCM output with data sets which are themselves generated by models. Palutikof *et al.* (1990), for example, compared near-surface wind speeds interpolated from GCM grid-point output with those estimated from a boundary-layer model initialized with observed geostrophic winds.

Inter-model comparisons and comparisons with observations tend to be qualitative rather than quantitative in nature. Statistical methods are, however, being developed (Karl *et al.*, 1990; Santer and Wigley, 1990; Wigley and Santer, 1990). The wider application of these methods should allow the *relative* reliability of individual models to be assessed.

Despite the technical problems of validation, as GCMs have improved so has their ability to reproduce the major features of the observed climate, such as the seasonal cycle. Nevertheless, the models still persistently fail to reproduce many of the details of present-day climate. Discrepancies in the predicted and observed strength and location of the Iceland Low and Azores High, for example, are commonly observed (Santer, 1988). Northwestern Europe, and the British Isles in particular, are very sensitive to atmospheric circulation changes related to shifts in the position and intensity of these systems.

4.4.5 Regional discrepancies between perturbed climates

It can be argued that the greatest confidence can be placed in those results from the perturbed run, relative to the control run, which are consistently duplicated in all model studies. However, it is possible that all models may produce the same persistent errors. Duplication does not necessarily imply that a result is correct. Relative confidence in a particular result will, however, be increased if it can be both duplicated and explained by a plausible physical mechanism. Enhanced high-latitude warming, for example, is a common feature of the perturbed run of equilibrium-mode GCMs. It can be explained by the removal of the ice–albedo feedback mechanism as melting occurs. On this basis, the relative degree of confidence that can be placed in various aspects of GCM results can be assessed (Table 4.4; Mitchell *et al.*, 1990).

Table 4.4 GCM results from equilibrium experiments for a doubling of atmospheric CO_2, and an estimate of the relative confidence that can be placed in these results.

Model results	Relative confidence
Global scale	
Warming of lower troposphere	High
Increased precipitation	High
Cooling of stratosphere	High
Zonal–regional scale	
Reduced sea ice	High
Enhanced Northern Hemisphere polar warming (especially in winter half-year)	High
More absolute high temperature extremes	High
Increased continental dryness	Moderate
Stronger monsoon	Moderate
Regional pattern in detail	Unknown
Greater/less interannual variability	Unknown
Precipitation extremes	Unknown

While all equilibrium-mode models concur that, with a doubling of CO_2, global temperature and precipitation will increase, and that the increases will be greatest at high latitudes, there are significant regional differences between the perturbed runs of GCMs. This is a particular problem so far as variables such as precipitation and soil moisture are concerned (Mitchell and Warrilow, 1987). The GFDL model, for example, predicts a 30–50% reduction in summer soil moisture over western Europe, but this summer drying is not apparent in the NCAR model simulations (Manabe and Wetherald, 1987; Meehl and Washington, 1988). In this case, the differences in the perturbed runs of the two models can be directly linked to differences in the control runs. Model discrepancies also arise from variations in the parametrization of climate and climate-related processes. Consensus between models tends to be particularly poor in sensitive regions such as northwestern Europe, where relatively minor circulation changes can have a major effect on climate.

Table 4.4 clearly demonstrates an important point concerning scale and model reliability. We have reasonable confidence in global and large-scale estimates of temperature and, to a lesser extent, precipitation (see, however, the discussion on transient models below). The relative level of confidence decreases as one moves from the global to the regional scale and when parameters other than mean temperature and precipitation are considered.

4.4.6 The transient climate response

Equilibrium models are unrealistic in terms of their representation of both enhanced greenhouse gas forcing and the climate response. Atmospheric concentrations of greenhouse gases will not instantaneously double but are increasing at a steady rate, currently estimated to be about 0.4% per annum (Bretherton *et al.*, 1990). The climate response to this transient forcing will be delayed by several decades because of ocean heat storage. Two types of transient model experiment can be identified: first, those in which atmospheric concentrations of greenhouse gases are increased incrementally, rather than instantaneously (transient forcing); and second, those in which the evolution of the climate response over time, including lags due to the thermal inertia of the oceans, is modelled (transient climate response). A model may be run in one, or both, of these modes. The transient climate response to either a step change in CO_2 concentration or to transient forcing may, therefore, be investigated.

Relatively simple energy balance and box diffusion-type models can be used to model the global and zonal transient response to enhanced greenhouse gas forcing. The IPCC, for example, recently assessed rates of warming estimated by a global energy balance upwelling-diffusion model, assuming a Business-as-Usual (BaU) emissions scenario. In this particular model experiment, both transient forcing and the transient climate response are represented. Over the period 1990–2030 the estimated global warming is 0.7–1.5°C (best estimate 1.1°C) and over the period 1990–2070 it is 1.6–3.5°C (best estimate 2.4°C). Members of the IPCC group have used these estimates of the mean global transient warming to scale down output from the perturbed $2 \times CO_2$ runs of a number of high-resolution equilibrium-mode GCMs (Mitchell *et al.*, 1990). Assuming the BaU emissions scenario, the equilibrium GCM output is scaled down to correspond to a mean global warming of 1.8°C. This is the actual warming (from pre-industrial times) estimated by the energy balance model to occur by 2030, relative to a final $2 \times CO_2$ equilibrium warming of 2.5°C. The climate changes in the GCMs are assumed to be linear in time.

In order to investigate the transient *regional* response, GCMs are required in which ocean heat transport and dynamics are fully incorporated. This has been done by coupling atmosphere and ocean GCMs. Coupled atmosphere/ocean (A/O) GCMs have a very large computing requirement and to date only a few fully coupled model runs have been performed (Harvey, 1989c; Bretherton *et al.*, 1990).

In addition to the large computing requirement, there are a number of technical problems associated with the use of transient A/O GCMs. Many of these relate to differences in the response time between the atmosphere (rapid) and the oceans (very slow). Coupled atmosphere and ocean GCMs tend to 'drift' out of radiative balance. Validation of coupled models is complicated by the lack of observations over the oceans. Where data are available, comparison reveals substantial control run errors. Washington and Meehl (1989) found, for example, that many ocean models display errors in the present-day SST simulations. Coupled A/O GCMs are at an early stage of development and confidence in their results remains low. It is, however, interesting to compare the climate response of these models to enhanced greenhouse gas forcing with that of equilibrium-mode models.

Washington and Meehl (1989) ran the NCAR GCM coupled with an ocean model using an instantaneous doubling of CO_2 (from 330 to 660 ppmv) and a transient forcing scenario (CO_2 increasing over 30 years at 1% per annum from 330 ppmv to 429 ppmv). In the first case (660 ppmv CO_2), the estimated mean global warming was 1.6°C. In the second, transient forcing, case (429 ppmv CO_2) the warming was only 0.7°C.

Although the large-scale pattern of temperature change in both equilibrium- and transient-mode runs tends to be broadly similar, major differences do occur. In Section 4.4.5 we concluded that, since there is a plausible explanation, we can be relatively confident in the GCM results that show enhanced greenhouse warming at higher latitudes. This feature is not, however, seen in recent A/O GCM experiments (Stouffer *et al.*, 1989; Washington and Meehl, 1989; Manabe *et al.*, 1990; 1991).

Stouffer *et al.* (1989) investigated the response of the GFDL atmospheric GCM coupled with a dynamic ocean model to a CO_2 increase of 1% per annum over a 100-year period. This forcing is equivalent to a doubling of CO_2 by 2030. A marked asymmetry was found in the response of the two hemispheres. Warming reduced towards high latitudes in the Southern Hemisphere and was very slow in the circumpolar oceans of the Southern Hemisphere. These results reflect the large thermal inertia and the vast extent of the Southern Hemisphere oceans. In the Northern Hemisphere, warming increased with latitude, with the notable exception of the northern North Atlantic. In this region warming was slow. By the seventh decade of the simulation, a 25% reduction in the strength of the thermohaline circulation of the northern North Atlantic was observed. The ocean circulation changes in this region are considered to be related to feedback mechanisms involving evaporation, salinity changes and thermal advection (Stouffer et al., 1989).

This study has been extended by Manabe et al. (1990; 1991). In the more recent study, three 100-year integrations were run. In run G, CO_2 was increased at a rate of 1% per annum. In run S, CO_2 was held constant, and in run D, CO_2 was decreased at a rate of 1% per annum. In order to offset the bias of the coupled A/O model, the fluxes of heat and water at the ocean–atmosphere interface were adjusted by a fixed amount each year, depending on season and geography. In order to allow the effects of ocean thermal inertia to be identified, the atmospheric model was also run in equilibrium mode combined with a 50 m mixed-layer ocean. Three equilibrium mode runs were performed: $1 \times CO_2$, $2 \times CO_2$ and $\frac{1}{2} \times CO_2$.

Over most of the Northern Hemisphere and at low latitudes in the Southern Hemisphere, the distribution of temperature change at the time of doubling/halving of CO_2 is similar in the transient and equilibrium models (Manabe et al., 1991). The transient response, however, is very slow over the northern North Atlantic and the circumpolar Southern Hemisphere oceans so that, at the time of CO_2 halving/doubling, over the northern North Atlantic the ratio of transient to equilibrium response is less than 0.4. In the immediate vicinity of the Antarctic continent, no warming occurs in the transient model. The slow response of the transient model is attributed to deep vertical mixing.

The thermal inertia effect is found to be greater during cooling (integration D) than during warming (integration G). In integration G, the increase in precipitation at high latitudes is larger than the evaporation change, hence surface salinity is reduced. The stability of the water column therefore increases, less deep mixing occurs and the surface temperature anomalies do not penetrate as deep. As the strength of the thermohaline circulation weakens, the northward advection of warm saline water is also reduced, further slowing the rate of warming.

The reverse occurs in the cooling experiment, integration D. Evaporation is reduced, surface salinity increases and the static stability of the water column is reduced at high latitudes. The strength of the thermohaline circulation is, therefore, increased. Analysis of ocean core records, however, indicates that the thermohaline circulation is actually weaker during glacial and stadial periods (see Section 2.4.3). In integration D, the cold water anomaly penetrates deeper than does the warm water anomaly in integration G. Although the thermal inertia effects are greater in integration D, the magnitude of the global surface air-temperature response is similar in experiments G and D (Manabe et al., 1991). This is because the larger thermal inertia effects are offset by positive ice–albedo feedbacks during cooling.

These results confirm the importance of the oceans in modifying the response of the climate system to enhanced greenhouse warming. However, a number of problems with the model can be identified. In order to couple successfully the atmospheric and ocean models, for example, it is necessary to apply surface flux adjustments which are as large as the fluxes themselves (Manabe et al., 1991). Also, the atmospheric model tends to overestimate precipitation at high latitudes in control run S.

Similar results for the Southern Hemisphere were found by Washington and Meehl (1989) using the NCAR GCM with transient CO_2 forcing (increasing at 1% per annum). In this study, however, the reduced warming effect over high-latitude oceans was found to be comparable in both hemispheres. A net cooling was observed in the zone 60–80°N. Again this effect was associated with a reduction in the strength of the thermohaline circulation.

Although transient models are still at the developmental stage, these results are disquieting. It seems possible that the equilibrium GCM results in which we have greatest relative confidence, such as enhanced high-latitude warming, are misleading.

The first results from transient models are interesting in what they can tell us about potential rates of climate change and the patterns of the time-dependent changes. Washington and Meehl (1989), for example, have compared the zonal pattern of change averaged over years 6–10 of transient CO_2 forcing (increasing at 1% per annum) with that averaged over years 26–30. In the early period (years 6–10), the greatest surface temperature increases occur near 30°N and 40°S. Minimum surface warming occurs at high latitudes. Warming of the oceans is confined to the top two layers of the model ocean. Weak cooling of the ocean surface layer occurs in a narrow band centred on 80°N. During the later period (years 26–30), maximum atmospheric and ocean warming occurs near 20°N and 45°S. Changes in ocean temperature are largely confined to the surface layer. No uniform atmospheric or ocean warming occurs in any latitude band. Cooling now occurs north of about 30°N, reaching a maximum near 65°N. The cooling is more extensive than when the same, transient climate response, model is run assuming an instantaneous doubling of CO_2 rather than an incremental increase (Washington and Meehl, 1989).

Transient-response A/O GCMs are currently still in the developmental stage. The very large computing requirement means that progress is slow. No substantial attempts to develop regional scenarios based on transient model output have been made. However, the results emerging from this new generation of models raise many complex and intriguing questions about the nature of climate change.

4.5 Conclusions

The major potential mechanism of climate change over the next few hundred years is considered to be the enhanced greenhouse effect. The length of time over which the warming effect will persist is, however, uncertain. This is due, in part, to uncertainties about the sources and sinks of greenhouse gases, the nature of the response and the influence of future policy changes. The major uncertainty, however, relates to the interaction between short-term greenhouse gas-induced forcing and longer-term orbital forcing. This issue is discussed further in Chapter 5.

Despite their many uncertainties and known errors, GCMs are considered to offer the greatest potential for developing regional scenarios of climate change for use in impact studies (Giorgi and Mearns, 1991; Grotch and MacCracken, 1991; see Section 9.2). Equilibrium-mode GCMs have been widely used in the construction of high-resolution regional scenarios. Increasingly sophisticated methodologies are being developed to address the problems associated with the low spatial resolution of the GCM grid-point output.

More fundamental problems remain, related to the validity of the model results themselves. Comparisons of control run results with present-day data suggest that GCMs are not always able to reproduce the present-day features of the global climate. Whereas considerable improvement has been achieved by increasing model resolution, doubts still remain concerning the representation of feedback effects. Until substantial advances in modelling occur, none of the available GCMs can provide reliable high-resolution scenarios suitable for use in quantitative regional impact studies. They may, however, be used to assess the sensitivity to climate change of, for example, agriculture, and in the development of impact assessment methodologies, under the condition that their limitations are fully acknowledged.

The results from fully coupled A/O GCMs, which incorporate the transient nature of greenhouse gas forcing and the associated warming, are becoming available. These suggest that the regional pattern of change differs markedly if time-dependent effects such as the thermal inertia of the oceans and the transient nature of the forcing are taken into account. The technical problems associated with these models are, however, great and it may be some time before they can be used in the construction of regional climate change scenarios.

In this chapter, discussion has been confined to the expected changes in standard climate parameters such as temperature and precipitation. Changes in sea level are also expected to accompany enhanced greenhouse gas warming and are considered in Chapter 6. First, however, we examine the relationship between short-term greenhouse gas-induced forcing and longer-term orbital forcing.

5 Natural versus anthropogenic forcing

5.1 Introduction

Over about the last 100 years the global surface air temperature has risen by 0.3–0.6°C (Folland *et al.*, 1990). Transient models predict an anthropogenic greenhouse gas-induced warming of 0.4–1.2°C over approximately the same period (Wigley and Barnett, 1990). The observed temperature changes, therefore, lie at the lower end of the model predictions.

If the greenhouse effect is real, then, from the model results one would expect the observed warming to have been somewhat greater. We know that greenhouse gas emissions have been rising since pre-industrial times, but this is not the only climate forcing mechanism in operation (Chapter 3). Mechanisms such as solar variability, volcanic eruptions and atmosphere–ocean feedbacks can all lead to climate changes over relatively short time scales. Such mechanisms may enhance or dampen the effect of enhanced greenhouse warming (Kerr, 1989b; Wigley and Kelly, 1990). In addition, the climate system demonstrates random variability, or noise, which makes the detection of greenhouse gas-induced warming difficult (Wigley and Barnett, 1990).

On short time scales, therefore, the climate changes induced by natural and anthropogenic forcing mechanisms interact. Interactions are also expected to occur on longer time scales, notably between the climate changes resulting from the enhanced greenhouse effect and orbital forcing (Mitchell, 1977; Goodess *et al.*, 1988; 1990).

5.2 Orbital versus greenhouse gas forcing

Given the different time scales of orbital and greenhouse gas forcing, there is no direct observational evidence which can be used as a basis for exploring the relationship between the two mechanisms. There is no alternative other than to rely on informed judgement and speculation, together with some very limited modelling evidence.

The simplest assumption that can be made about the relationship between orbital and enhanced greenhouse gas forcing is that a relatively short episode of greenhouse gas-induced warming will be followed by a switch back into the 'natural pattern' of glacial–interglacial cycles. This implies that greenhouse gas emissions and global temperature will eventually return to levels close to those of the present day. Mitchell (1977), for example, interposes a greenhouse gas-induced 'super-interglacial' between the end of the present interglacial and the cooling trend leading towards the next glacial episode. He postulates that such a super-interglacial episode could delay the onset of cooling by 2 ka. A similar approach has been used to develop a climate index representing the succession of major climate states likely to be experienced in the UK over the next 1 Ma (Goodess *et al.*, 1988; 1990). In this index, greenhouse gas-induced warming is represented by a 1 ka super-interglacial occurring between the present day and the start of oscillatory cooling leading to the next glaciation about 60 ka AP. How realistic is the assumption that greenhouse gas-induced warming will persist for 1–2 ka?

In Chapter 4, a distinction was made between the equilibrium and the transient response to greenhouse gas forcing. Because of the thermal inertia of the oceans, lags appear in the response of the climate system to forcing. Even if all anthropogenic greenhouse gas emissions were to cease tomorrow, the world would continue to warm for several decades. (This concept of 'committed' change is also discussed in relation to sea-level rise in Chapter 6.)

In order to obtain an upper limit of warming due to CO_2, it could be assumed that all fossil fuel reserves will eventually be burnt. However, questions such as the rate at which they are used and whether or not the exploitation of potential reserves is technically feasible or economically viable, still arise. The IPCC expert group did not attempt to estimate emissions beyond 2100 (Shine *et al.*, 1990). This reflects the difficulties involved in assessing future technological developments and energy and environmental policy. It is equally difficult to make long-term predictions for the other greenhouse gases. It is not, for example, a trivial task to judge how successful recent attempts at reducing CFC use will be, or the extent to which CFCs will be replaced by other radiatively active gases. In order to investigate long-term methane emissions, the extent of wetlands, among other factors, needs to be assessed. Their area is likely to change as a result of greenhouse gas-induced warming but, given the shortcomings of the climate models discussed in Chapter 4, it is difficult to make reliable predictions of regional ecosystem changes.

Even under present-day conditions, the roles of the oceans and the biosphere within the global carbon cycle are not fully understood. The atmospheric CO_2 concentration at any time reflects the balance between emissions and uptake by the oceans and biosphere. The saturation points of these sinks and their behaviour under greatly increased CO_2 concentrations are unknown. We know that atmospheric CO_2 concentrations increase during interglacial periods and decrease during glacial periods (see Chapter 2). However, even at the peak of the last interglacial they only reached about 75% of present-day concentrations. It is necessary to go back millions of years through the geological record, to the Cretaceous period, to find significantly higher concentrations. Thus, there are no suitable analogues for studying the global carbon cycle in a high-CO_2 world.

So far, we have only considered the rate of increase in atmospheric concentrations of greenhouse gases. Eventually these concentrations will start to decrease. The somewhat simplistic representation of enhanced greenhouse warming as an intervening super-interglacial in a 'normal' glacial–interglacial cycle (Mitchell, 1977; Goodess *et al.*, 1988; 1990) assumes that greenhouse gas emissions will revert to present-day or pre-present levels and that global temperatures will follow suit. Given the present state of knowledge, however, it is not possible to make any reliable estimate of the time it may take to return to, for example, pre-industrial atmospheric CO_2 concentrations.

Furthermore, it is possible that mean global surface temperature may stabilize at a level significantly higher than the present day. We may even visualize a society with advanced scientific and technological capabilities which decides deliberately to adjust emissions in order to suit its perceived needs. At the furthest extreme, for example, it could be argued that society might positively encourage greenhouse warming in an attempt to avert a future glaciation. However, we consider that it is unwise to speculate about the future behaviour and capabilities of society and that it is reasonable to assume greenhouse gas emissions and global temperature will eventually return to levels close to those of the present day.

The second assumption which needs to be examined is that only the length and not the strength or impacts of enhanced greenhouse warming will influence the future operation of orbital forcing mechanisms.

5.3 Long-term effects of a greenhouse gas-induced warming

Enhanced greenhouse warming is expected to cause changes beyond its impact on the atmospheric circulation. These changes will, in turn, have implications for the climatic response to orbital forcing. Changes in the cryosphere, such as the retreat of small glaciers and the Greenland ice sheet, are likely to be critical. The positive feedback mechanisms which help to transform the relatively weak insolation changes associated with orbital forcing into global glacial–interglacial cycles are discussed in detail in Section 2.4. The cryosphere and the polar ice sheets are considered to play a particularly important role through isostatic adjustment, albedo and/or general circulation effects.

It is conceivable that the impact of greenhouse warming on global cryospheric boundary conditions may seriously limit the effectiveness of these feedback mechanisms. At one extreme, it might even be imagined that they could be weakened to such an extent that the initiation of further glaciation would be prevented. Although the Quaternary pattern of glacial–interglacial cycles would not be repeated, some form of climate cycling on the same time scale would be expected. Geological evidence from pre-Quaternary periods, for example, indicates that sedimentary cycles with orbital periodicities occurred (see Chapter 2). They were not, however, accompanied by the advance and retreat of ice sheets.

Ideally, the sensitivity of orbital forcing to changes in cryospheric boundary conditions would be tested using a physically realistic model of the time-dependent behaviour of the coupled climate system including the cryosphere, atmosphere, oceans, lithosphere and biosphere. As discussed in Section 2.3, the development of such a model is constrained by computing power and speed. The model which comes closest to meeting these criteria is the sectorially averaged time-dependent physical climate model being

developed by Berger and colleagues (Berger *et al.*, 1991b). The structure of
this model and simulations of the last glacial–interglacial cycle were outlined
in Section 2.3. The model has also been used to test the sensitivity of orbital
forcing to potential anthropogenic climate change (Berger *et al.*, 1991b).

In this sensitivity experiment, it is assumed that greenhouse warming will
melt the Greenland ice sheet totally. Thus, the model is run with no
Greenland ice sheet and present-day CO_2 concentrations. The perturbed
and control simulations for the next 80 ka are shown in Figure 5.1. The main
results are that ice sheets do not reappear in the Northern Hemisphere until
about 15 ka AP, and that the next glaciation is delayed by a few thousand
years and is less extensive (Berger *et al.*, 1991b).

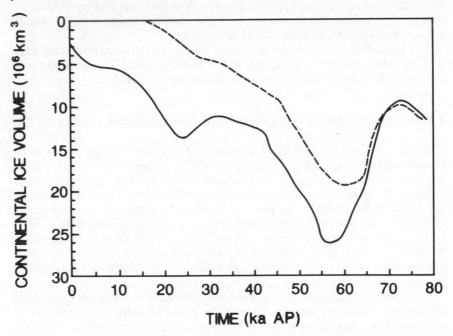

Figure 5.1 Future continental ice volume simulated for two different initial states:
with the Greenland ice sheet (solid curve) and without the Greenland ice sheet
(dashed curve). (From Berger *et al.*, 1991b.)

The representation of greenhouse warming in this experiment is
somewhat crude and the model does not, as yet, have a carbon cycle. Here
we assess the reliability of the experiment as a guide to the sensitivity of
orbital forcing effects to potential changes in cryospheric boundary
conditions. In particular, we consider evidence concerning the sensitivity of
the Greenland ice sheet to orbital forcing and enhanced greenhouse
warming.

It has been suggested that the Greenland ice sheet may have disappeared
during the last interglacial maximum at about 125 ka BP (Koerner, 1989).
However, the evidence on which this hypothesis is based has been strongly

criticized (Souchez *et al.*, 1990). Recent modelling evidence also indicates that the Greenland ice sheet has been, and will remain, a fairly robust feature of the cryosphere.

A three-dimensional time-dependent thermomechanical model with a fine resolution grid has been used in a series of experiments to investigate the sensitivity of the Greenland ice sheet to future warming (Huybrechts *et al.*, 1991; Letréguilly *et al.*, 1991). The major model inputs are bed topography, surface temperature, initial mass balance and thermal parameters. The response of the underlying bedrock to the changing ice load is incorporated.

In the first set of experiments, the ice-sheet response to step changes in temperature of 0–10°C is investigated (Letréguilly *et al.*, 1991). For warming of up to 2°C, the effects are hardly noticeable and, for warming of up to 3°C, they can only be detected at the ice-sheet margins, particularly in the southwest. With a warming of 4°C, the ice sheet splits into two parts: a larger ice sheet covering northern and central Greenland, with a smaller ice sheet in the south. With a 6°C temperature increase, only four small ice caps remain in the southern and eastern mountains 20 ka after the step change and, with an increase of 8°C, all remaining ice masses disappear about 5 ka after the step change. In this first set of experiments, the ice sheet generally attains equilibrium within 30 ka. Since the last interglacial (the Ipswichian, centred on 125 ka BP) only lasted for about 10 ka, and is considered to have been accompanied by a maximum global temperature increase of about 2°C above the present value, these model results imply that the Greenland ice sheet did survive (Letréguilly *et al.*, 1991).

In the second set of experiments, it is assumed that the entire Greenland ice sheet has melted (Letréguilly *et al.*, 1991). The model run commences with bare bed-rock (after isostatic uplift) and the effects of temperature increases of 0–6°C are investigated. With no temperature increase, a very similar ice sheet reforms. With warming of 2°C, it reforms more or less completely. Even with a warming of 3°C, two small ice sheets form, but equilibrium is not reached until after 50 ka. These results indicate that the Greenland ice sheet is not a 'relict' ice sheet owing its existence to former colder conditions (Letréguilly *et al.*, 1991). Even if it does melt during a future period of enhanced greenhouse warming, it is likely to reappear during a period of cooler climate.

Finally, in the third set of experiments, the response of the Greenland ice sheet to the Villach II enhanced greenhouse warming temperature scenario (Jäger, 1988) is assessed (Huybrechts *et al.*, 1991). The results indicate that, for global warming of less than 2.7°C, ice accumulation is greater than runoff. However, for a global warming of 4.2°C by 2100, some melting occurs and the ice sheet settles to a new equilibrium at about half its present volume. For extreme warming (over 8°C by 2100), complete melt occurs in about 500 a.

The potential magnitude and rate of enhanced greenhouse warming is discussed in Chapter 4. The best estimates of equilibrium global temperature rise associated with a doubling of the equivalent CO_2 concentration lie in the range 1.5–4.5°C (MacCracken and Luther, 1985;

Bolin *et al.*, 1986). The estimated transient warming over the period 1990–2070, given a BaU emissions scenario, is 1.6–3.5°C, with a best estimate of 2.4°C (Bretherton *et al.*, 1990). Equilibrium GCMs suggest that warming will be enhanced at high latitudes. However, emerging results from coupled A/O transient GCMs suggest that warming may actually be at a minimum at high latitudes, and that cooling may occur in the northern North Atlantic (Stouffer *et al.*, 1989; Washington and Meehl, 1989; Manabe *et al.*, 1991). Thus we conclude that it is unlikely that the Greenland ice sheet will totally disappear during the projected period of enhanced greenhouse warming.

If the Greenland ice sheet is not expected to disappear, how valid is the Berger *et al.* (1991b) model study? Unless cooling occurs in the northern North Atlantic, as predicted by some transient, coupled A/O GCMs (see Section 4.4.6), some reduction, possibly by up to 50%, in the volume of the Greenland ice sheet is expected to accompany warming. In addition, a general reduction in the extent of Northern Hemisphere sea ice, mountain glaciers and snow cover will occur (Barry, 1985; Houghton *et al.*, 1990). It is expected that the reduction in snow cover will lead to a lower surface albedo, thus more solar radiation will be absorbed at the surface. This is a positive feedback effect. However, analysis of snow–climate feedback mechanisms in 17 GCMs suggests that additional amplification or moderation may be introduced by cloud interactions with long-wave radiation (Cess *et al.*, 1991; see discussion in Section 4.4.3). Comparison of output from these GCMs indicates that the net result of snow–climate feedbacks ranges from a weak negative to a strong positive effect (Cess *et al.*, 1991). Finally, we note the accumulation of modelling evidence indicating that the mass balance of the Antarctic ice sheet is likely to remain positive in a high-greenhouse-gas world (Huybrechts and Oerlemans, 1990; Warrick and Oerlemans, 1990; see discussion in Chapter 6).

Overall, the available evidence suggests that the cryosphere will be reduced in a high-greenhouse-gas world. We conclude that, for the purposes of a sensitivity experiment in which we are primarily interested in changes in albedo, it is not unreasonable to represent this reduction by the absence of the Greenland ice sheet. Furthermore, the modelling evidence discussed above supports the re-establishment of ice sheets, albeit to a reduced extent initially, following an episode of enhanced greenhouse warming.

The limited modelling evidence which is available suggests that enhanced greenhouse warming will delay the onset of the next glaciation and will slow and restrict the development of the major ice sheets. The next glaciation is likely to be less severe than would have been the case in the absence of the enhanced greenhouse effect. The switch back into glacial–interglacial cycling is expected to be gradual. However, an alternative view has been proposed. The possibility of abrupt mode changes in the thermohaline circulation and, hence, in regional climate conditions (Broecker, 1987; Broecker and Denton, 1989; 1990) is discussed in earlier chapters. At the present time, however, there is no modelling evidence available which allows assessment of the possibility of a mode-switch transition (that is, a sudden jump) between anthropogenic and orbital forcing.

5.4 Conclusions

There are clearly great uncertainties surrounding the relationship between greenhouse and orbital forcing, and much scope for speculation. Three possible patterns describing the relationship between enhanced greenhouse warming and orbital forcing can be identified (Goodess *et al.*, 1992). First, the simplest assumption is that a relatively short (say, 1 ka) period of greenhouse gas-induced warming will be followed by a switch back into the 'natural pattern' of glacial–interglacial cycles. The second possibility is that, following a period of warming, the next glaciation will be delayed and will be less severe. The third possibility is that enhanced greenhouse warming will so weaken the positive feedback mechanisms which transform the relatively weak orbital forcing into global interglacial–glacial cycles, that the initiation of future glaciations will be prevented. This is the so-called 'irreversible greenhouse effect'.

On the basis of the limited evidence available, we consider that the second possibility, delayed and reduced glaciation, has the highest probability. In particular, the most likely effect of greenhouse gas-induced warming will be to delay the onset of the next glaciation by about 5 ka, from 55 ka to 60 ka AP, and to reduce its severity. The transition to cooling is likely to be delayed for more than 1 ka. Although the transition is expected to be rapid, an abrupt mode-switch of the type proposed by Broecker is considered unlikely. We assume that the climate system will recover from the effects of enhanced greenhouse warming and that orbital forcing will, therefore, operate much as it has over the Quaternary. Based on these assumptions, it is reasonable to use the geological record of the Quaternary period as a guide to the future, beyond the point of transition. We consider the possibility that an 'irreversible greenhouse effect' will occur to be remote.

Finally, we note that greenhouse gas-induced warming is the only form of anthropogenic forcing considered in this book. Over very long future time scales, however, the possibility exists of additional, as yet unrecognized, anthropogenic forcing mechanisms.

6 Sea-level change

6.1 Introduction

Changes in sea level are known to accompany glacial–interglacial cycles. Scientific reasoning suggests that we can expect enhanced greenhouse warming to cause sea-level changes in the short-term future. In this chapter, we briefly review the major mechanisms. The records of glacial–interglacial and post-glacial sea-level change provide a guide to future long-term changes. Similarly, analysis of sea-level change over the last 100 a permits identification of the major contributors to future greenhouse gas-induced sea-level rise. Finally, we discuss estimates, by a number of different groups, of the magnitude of sea-level changes over the next century.

6.2 Mechanisms

Mechanisms of sea-level change operate over a number of different spatial and temporal scales (Van der Veen, 1988; Warrick and Oerlemans, 1990; Clayton, 1991). Principal mechanisms are summarized in Table 6.1. The relationships between the various processes are illustrated in Figure 6.1. This diagram is highly simplified and does not include feedback mechanisms (Van der Veen, 1988).

Given the problems of defining a fixed reference or base level, sea-level changes are generally discussed in terms of the *relative* variations. Relative sea level at any particular location represents the balance between global and local mechanisms, on the one hand, and between long-term and short-term forcing, on the other. In this chapter, we are principally concerned with global changes in sea level. Greater consideration is, for example, given to eustatic change associated with variations in global ice volume than to the more localized effects of isostatic adjustment due to ice-sheet loading and unloading.

6.3 The record of sea-level change

6.3.1 Glacial–interglacial sea-level change

Present-day global temperatures are relatively high in comparison with those experienced over the course of the Quaternary period. Global sea level is also relatively high, thus obscuring much of the direct geological evidence of past fluctuations.

Coral reef terraces which, because of tectonic uplift, lie above present-day sea level, are one source of evidence on past conditions (Mesolella *et al.*,

Table 6.1 Mechanisms of sea-level change.

---

1. Eustatic change
The major processes contributing to eustatic sea-level change are glacio-eustatic movements in sea level due to alternating glaciation and deglaciation. Glacio-eustatic sea-level fall occurs when vast volumes of water are locked up in the cryosphere during glacial periods. Sea-level rise occurs with the melting of the cryosphere during periods of deglaciation. Thermal expansion and contraction of near-surface water also contribute to eustatic sea-level change to a limited degree.

2. Glacial isostasy
The alternate loading and unloading of the Earth's surface due to the presence and subsequent removal of large ice sheets causes sea-level fluctuations in the vicinity of areas subjected to glaciation.

3. Long-term tectonic change
Tectonic processes such as continental drift and mountain-building cause changes in sea level. Subsidence of the Earth's surface causes total sea area to increase and sea level to fall globally. However, locally, subsidence will cause relative sea level to increase. Subsidence itself may be caused by a number of tectonic processes, including sediment loading and the cooling of crustal belts following volcanic activity, rifting and earthquakes.

4. Hydro-Isostasy
Glacial isostasy is a well accepted mechanism. The alternate loading and unloading of the ocean floors by different depths of water has, in contrast, received little attention.

5. Geoidal changes
The equipotential surface of the Earth's geoid is controlled by the interrelationship of the Earth's gravity and shape. The pressure exerted varies around the globe and is considered to result in sea-level differences of several metres between different parts of the globe. It has also been suggested that changes in the geoid may occur with glacial–interglacial fluctuations causing a latitudinal redistribution of global sea level.

6. Variations due to climatic fluctuations
Changes in oceanic circulation, including El Niño-Southern Oscillation effects, can influence sea level. The Earth's rotational deflective force, known as the Coriolis force, can also influence sea level, particularly along higher latitude, eastern coasts in the Northern Hemisphere.

Sources: Goodess *et al.*, 1988; 1992.

1969; Bloom *et al.*, 1974; Fairbanks, 1989; Bard *et al.*, 1990; Ku *et al.*, 1990). The reefs develop when sea level rises faster than the land, and are preserved when tectonic uplift continues. Coral reef terraces indicate relatively stable periods of high sea level. In this respect they are comparable with raised beaches. Most studies are based on reefs from Barbados and New Guinea: these regions are subject to ongoing and steady tectonic uplift. By their very nature, coral reef records are discontinuous, recording only sea-level maxima (indicated by terraces) and minima (indicated by reef geometry and evidence of deltaic environments).

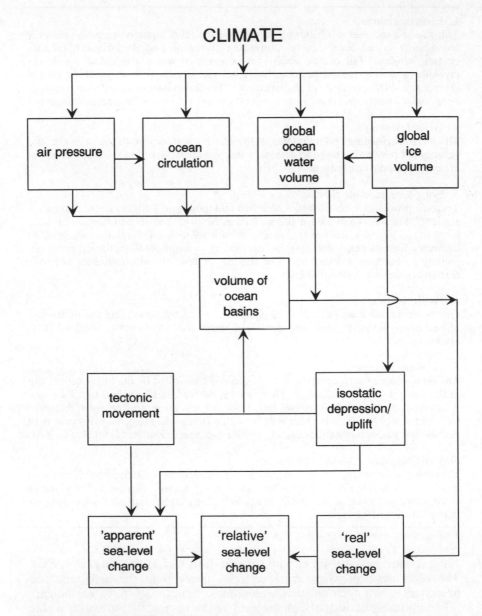

Figure 6.1 Relationships between the causes of sea-level change (highly simplified without feedback mechanisms). (After Van der Veen, 1988.)

Chappell and Shackleton (1986) have interpolated a continuous record of sea level for the last 300 ka from a series of raised coral terraces along the northeast coast of the Huon Peninsula in New Guinea. The rate of tectonic uplift is assumed to be constant over time, but varying along the coast from 0.9 m ka^{-1} to 3.5 m ka^{-1}. The terraces are dated by ^{14}C and ^{230}Th/^{234}U methods. The final sea-level time series is adjusted to fit the orbitally tuned SPECMAP time scale (Imbrie *et al.*, 1984). The agreement between estimates of high sea-level stands from the Huon Peninsula and from other regions such as Barbados, Timor and Vanuatu is relatively good (see Chappell and Shackleton, 1986, Table 1). This suggests that the record is representative of global trends.

Chappell and Shackleton consider that their adjusted record for the last 130 ka represents the best available evidence for the New Guinea/Australia region. It is shown in Figure 6.2, together with the oxygen isotope record from the Pacific Ocean core V19-30 (Shackleton *et al.*, 1983). In comparison with the present day, Figure 6.2 indicates that sea level has ranged from –130 m at glacial maxima (18 ka BP and 150 ka BP) to +6 m at interglacial maxima (132 ka BP). The apparently good agreement between the two records largely reflects the tuning of the sea-level record to the oxygen isotope record and to the SPECMAP time scale. It is concluded that, in fact, there is no systematic, linear, relationship between the oxygen isotope and sea-level records (Chappell and Shackleton, 1986).

The coral reef records of sea level are subject to a number of uncertainties, associated mainly with past rates of tectonic uplift and the reliability of dating methods. There is, for example, some uncertainty as to the number of peaks of high sea level associated with the last interglacial (stage 5e). Dates of raised terraces from Barbados have recently been reassessed using the high-resolution uranium-series method (Ku *et al.*, 1990). Dates for the peak of the last interglacial of 120 ± 2 ka BP and 117 ± 3 ka BP were obtained from two sites in Barbados (the Maxwell and Rendevous Hill reefs). Given the errors involved, Ku *et al.* conclude that these two dates are effectively the same, and that there is no evidence of multiple high sea-level stands during stage 5e. A similar conclusion is reached by Bard *et al.* (1990), using U–Th techniques to date Barbados corals.

New mass-spectrometric techniques for U–Th dating have recently been applied to 37 samples from two fossil coral reefs in the Bahamas (Chen *et al.*, 1991). The interglacial high sea-level stand, of about 6 m above present mean sea level, is estimated to have been achieved by 132 ka BP; it had certainly been reached by 129 ka BP. It was sustained until about 120 ka BP, when sea-level fell rapidly at a rate of about 20 m ka^{-1}. There is no evidence of a double interglacial peak in this study.

Speleothem deposits in coastal caves have been proposed as an additional proxy indicator of long-term sea-level change (Li *et al.*, 1989). Speleothem growth (i.e. calcite deposition) can only occur in air-filled caves. When sea level rises, coastal caves become flooded. During high sea-level stands, therefore, speleothem growth cannot occur. If speleothem deposits can be accurately dated at frequent intervals throughout a profile, it should be

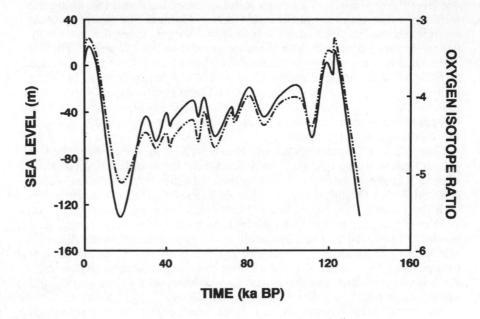

Figure 6.2 Estimates of global sea-level change over the last 140 ka derived from New Guinea coral reef records (solid curve) compared with the oxygen isotope record of global ice volume from ocean core V19–30 (dashed curve). Sea level is shown as departure (m) from present day. (Data from Chappell and Shackleton, 1986.)

possible to identify periods of high sea level (indicated by hiatuses in deposition and by the onset of erosion), periods of low sea level (indicated by deposition), and intermediate periods (indicated by the resumption of deposition). Thus, it is argued, the record of speleothem growth in coastal caves may be used as an indicator of sea-level fluctuations over a wider range than can coral reef terraces. A study has been made of flowstone deposits from 15 m below present-day sea level in a coastal cave in the Bahamas (Li *et al.*, 1989). The deposits are dated by the high-resolution mass-spectrometry uranium-series method. Evidence of relatively high sea level, similar to that of the present day, is apparent for six events over the 280 ka record: before 280 ka BP, 235–230 ka BP, 220–212 ka BP, 133–110 ka BP, 100–97 ka BP and after 39 ka BP. The most recent part of the record is lost because of erosion. These high sea-level stand estimates show general agreement with other sea-level records, and also with warm stages identified in marine oxygen isotope stratigraphies. The authors, therefore, conclude that speleothem deposits in coastal caves are a potentially useful proxy indicator.

One of the major problems with the use of coral reef and speleothem records for the reconstruction of precise histories of Quaternary sea level has been the requirement for accurate high-resolution dating methods. The studies discussed above (Chappell and Shackleton, 1986; Li *et al.*, 1989; Bard *et al.*, 1990; Ku *et al.*, 1990; Chen *et al.*, 1991) demonstrate some of the methodological advances which have been made in recent years. Chappell and Shackleton (1986) avoid some of the dating problems by adjusting their sea-level time series to fit the orbitally tuned SPECMAP time scale. Other studies (Li *et al.*, 1989; Ku *et al.*, 1990; Chen *et al.*, 1991) have made use of new high-resolution dating methods such as the mass-spectrometric techniques for U–Th and uranium-series dating. These recent studies provide more reliable results than do earlier studies, such as those of Mesolella *et al.* (1969) and Bloom *et al.* (1974).

A further problem with the use of coral reef and speleothem records as proxy indicators of sea-level change is that neither provides a continuous record. It is not appropriate to interpolate between sea-level maxima and minima on the basis of linear trends. A method for estimating sea level continuously through time from disjointed records has, however, been developed (Pinter and Gardner, 1989). Lagrange polynomial interpolation is used to fit a series of fourth-order equations to a record of sea level, in this case the Huon Peninsula, New Guinea, record of the last 140 ka (Bloom *et al.*, 1974). The Huon record consists of discrete data points representing high sea-level stands (identified from coral reef terraces) and sea-level minima (identified from deltaic gravel formations). The polynomial model developed by Pinter and Gardner (1989) is based on the statistical characteristics of high-resolution reconstructions of Holocene sea level (Fairbridge, 1961; Shepard, 1963; Mörner, 1969; Bloom, 1970). These Holocene records are shown in Figure 6.3 and are discussed in Section 6.3.2. The method adopted by Pinter and Gardner assumes that the pattern of sea-level change during previous post-glacial periods was essentially the same as that which occurred over the Holocene period. In addition, the model incorporates some general hypotheses about eustatic sea-level variations, together with an error-analysis procedure which accounts for errors in the original Huon data set. Errors may occur in both the dating and in the theodolite measurements.

The studies described above are all concerned with eustatic sea-level changes associated with variations in global ice volume, i.e. the amount of water locked up in the world's major ice sheets. Other studies have explored the relationships between sea level and the contributions from individual ice sheets. Studies relating to the Barents Sea, Greenland and the West Antarctic ice sheets are discussed below (Labeyrie *et al.*, 1986; Peltier, 1988b; Koerner, 1989).

The melting of large ice sheets leads to a redistribution of surface mass: the localized ice load is transformed into a more evenly distributed water load. Such a redistribution may have substantial dynamic effects on the Earth's rotation (Peltier, 1988b). The principal effects are a 'non-tidal' acceleration of the Earth's axial rate of rotation and modifications to the drift of the rotation pole with respect to the surface geography (i.e. polar wander). If it is assumed that 50–60% of the global sea-level rise indicated by tide-gauge records over the last 100 years or so is attributable to thermal

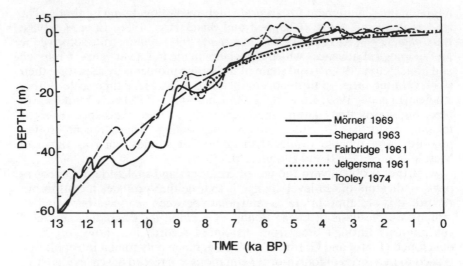

Figure 6.3 Post-glacial sea-level curves from Sweden (Mörner, 1969), Netherlands (Jelgersma, 1961), Gulf of Mexico (Shepard, 1963), northwest England (Tooley, 1974) and the world (Fairbridge, 1961). (From Tooley, 1974.)

expansion, the remainder must be due to melting of the Antarctic and Greenland ice sheets and/or of smaller ice sheets and glaciers (Peltier, 1988b). Peltier argues that substantial dynamic effects have accompanied ice sheet/glacier melt. He has tested this theory using a model of the Earth's rotational response to ice-sheet accretion and disintegration, developed using satellite observations of the Earth's rotation and tide-gauge sea-level records. The model is used to investigate the effect of Quaternary glacial–interglacial cycles on the Earth's rotation and on sea level. Initially, the model underestimates the predicted 'non-tidal' acceleration during the last deglaciation. In order to correct this apparent discrepancy Peltier invokes a large ice sheet centred near the Earth's rotation pole during the last glaciation. He concludes that the only possible location for such an ice sheet is the Barents Sea. However, the palaeoclimate evidence in support of a Barents Sea ice sheet is controversial. It is possible that the lack of agreement between the data and Peltier's results may be related to errors in previous estimates for the time of onset of deglaciation (Bard *et al.*, 1990; Lehman *et al.*, 1991). New high-resolution sea-level data from Barbados, for example, indicate that deglaciation may have started about 3 ka earlier than previously thought (Bard *et al.*, 1990).

It has been argued that debris in the basal ice of cores from the Canadian Arctic Islands and from Camp Century and Dye-3 in Greenland suggests that the Greenland ice sheet melted extensively, possibly completely, during the last interglacial (Koerner, 1989). Koerner considers that the ice texture and composition, including enrichment in the ^{18}O isotope, support the contention that basal ice is 'superimposed' ice typical of the early stages of ice-sheet development. Based on estimates of the Greenland ice-sheet extent and thickness, Koerner concludes that the severe reduction, or total disappearance, of the Greenland ice sheet would account for virtually all the 6 m enhancement in sea level relative to the present day, estimated for the last interglacial from coral reefs (Chappell and Shackleton, 1986). Thus, he concludes, it is extremely unlikely that the West Antarctic ice sheet disintegrated or was subject to strong surges in the last interglacial. Koerner's interpretation of basal ice has, however, been questioned (Souchez et al., 1990). For example, it is likely that the ^{18}O enrichment identified by Koerner is related to isotope exchange with hydroxyl-bearing minerals. Souchez et al. conclude that the available evidence in support of extensive melting of either the Greenland or the West Antarctic ice sheet in the last interglacial is controversial. In addition, the modelling evidence discussed in Section 5.3 does not support the total disappearance of the Greenland ice sheet during the last interglacial.

Although there is no evidence from West Antarctica of surges during the last interglacial, there is evidence of surges and associated calving in the West Antarctic ice shelves from the last glacial (Labeyrie et al., 1986). Comparison of planktonic and benthic foraminiferal records suggests that, between 35 ka and 17 ka BP, the Southern Ocean polar front was covered by a 'meltwater lid' largely derived from melting icebergs calved from the Antarctic ice shelves. It is argued that the most likely cause of this calving is a succession of surges of the ice shelves associated with ice-stream flow (Labeyrie et al., 1986).

From the above discussion, it is evident that knowledge of long-term changes in sea level is gradually improving. A reliable record of relative global sea level over the entire Quaternary period is not, however, available. It might be thought that such a record could be extrapolated from ocean core oxygen isotope records. Unfortunately, because of a number of interpretational problems, no simple linear relationship can be identified between the magnitude of ice volume, as indicated by ocean core records, and sea-level (Chappell and Shackleton, 1986; Bard et al., 1989; Li et al., 1989). These problems include difficulties in separating out the ice volume, temperature and sea-level signals in the ocean core records, together with dating uncertainties and problems of bioturbation (Chappell and Shackleton, 1986; Bard et al., 1989). It may, however, be possible to estimate sea level from ocean core records over shorter periods, such as the post-glacial period. Sea-level changes over this latter period are considered in Section 6.3.2.

6.3.2 Post-glacial sea-level change

Global

Over the post-glacial period, global sea level has risen. A number of sea-level curves for this period are shown in Figure 6.3. These curves are taken from relatively early studies, and have been refined by later research discussed below. Data come from northwest England (Tooley, 1974), Sweden (Mörner, 1969), the Netherlands (Jelgersma, 1961) and Mexico

(Shepard, 1963). A global curve is also shown (Fairbridge, 1961). The estimates are based on radiocarbon-dated shoreline indicators in the geological record such as molluscs, corals and brackish-water peat.

This group of curves demonstrates the influence of both global and local mechanisms on sea-level change. A rapid sea-level rise, of the order of 1 m per century, is seen in the earlier part of the record. This is attributed to the rapid melting of the Laurentide and Fennoscandinavian ice sheets following the Last Glacial Maximum. From about 6 ka BP onwards, the rate of increase is reduced. Estimates for the rate of increase over the last millennium are in the range of 1–2 cm per century (Gornitz et al., 1982; Klige, 1982; Robin, 1986). This trend is considered to reflect the rate of Antarctic and Greenland ice-sheet melting. The curves shown in Figure 6.3 can be divided into those which indicate a smooth long-term trend (Jelgersma, 1961; Shepard, 1963) and those indicating shorter-term oscillations (Fairbridge, 1961; Mörner, 1969; Tooley, 1974). In part, these differences are a function of the different methods of data analysis. However, it is reasonable to assume that the shorter-term fluctuations are a genuine feature of sea-level behaviour, although their magnitude is uncertain.

The main features of the post-glacial sea-level record are considered in more detail in the rest of this section, together with some more recent and refined estimates of the rate of change.

The rate of sea-level rise during the last deglaciation has recently become a topic of interest because of the light it may throw on the mechanisms of deglaciation and the causes of the Younger Dryas event (see Chapters 2 and 3). Bard et al. (1989) use coupled measurements of $\delta^{18}O$ and $\delta^{14}C$ from a particular species of planktonic foraminifera to make preliminary estimates of sea-level changes during the last deglaciation. The foraminifera records come from two North Atlantic cores (SU81-18 and SU81-14) which are located at about 37°N 10°W, well away from such major areas of meltwater discharge as the Gulf of Mexico and Gulf of St Lawrence. The estimates are obtained by subtracting a local temperature signal from the plankton record and establishing a linear relationship between the resulting ice volume record and sea level.

From their preliminary study, Bard et al. (1989) conclude that sea level rose by about 40 m over the period 14.5–13.5 ka BP, i.e. by a rapid rate of 4 m per century. Results from core SU81-18 indicate that by 12.2 ka BP, sea level was 67 ± 7 m lower than today and, at 8.2 ka BP, it was 24 ± 8 m lower. Comparable estimates are obtained for core SU81-14: 83 ± 10 m lower at 12.2 ka BP and 13 ± 11 m lower at 8.5 ka BP. These estimates are considered to agree reasonably well with independent results from coral reefs and other sources (Bard et al., 1989).

A continuous and detailed record of sea-level change since the last glaciation has been developed using results from a recent drilling program in Barbados (Fairbanks, 1989). Nine cores were taken through the coral reefs and subjected to radiocarbon dating. Sea-level changes over the last 17 ka were extrapolated from these dated stratigraphic records. Sea level at the Last Glacial Maximum is estimated to have been 121 ± 5 m below present

(Fairbanks, 1989). It is concluded that sea-level rise during deglaciation was not monotonic, but included two separate periods of very rapid rise. Over the period 17.1–12.5 ka BP it is estimated that sea level rose 20 m (a rate of about 0.4 m per century). After 12 ka BP sea level rose by 24 m in less than 1 ka (a rate of at least 2.4 m per century). This event is termed meltwater Pulse 1A and is associated with, but not exactly synchronous with, Termination 1A in the marine oxygen isotope record. The rate of rise then slowed considerably, reaching a minimum at the start of the Younger Dryas (about 11 ka BP). Following the Younger Dryas, the rate of rise increased over the period 10.5–10 ka BP. A second Meltwater Pulse, 1B, is identified, centred at about 9.5 ka BP. This is associated with Termination 1B in the marine oxygen isotope record.

The existence of two meltwater peaks is confirmed by Bard et al. (1990). U–Th dates obtained by mass spectrometry in corals from Barbados indicate two peaks at about 14 ka and 11 ka BP. Bard et al. consider that the U–Th dates are more precise than the ^{14}C dates obtained by Fairbanks (1989). It is likely that the meltwater surges exceeded the rate of 10^6 km^3 of continental ice per century (Bard et al., 1990).

Thus, the evidence indicates that the rate of sea-level change during deglaciation has not been constant (Table 6.2). Periods of very rapid rise (in the order of 2–4 m per century) alternate with periods of much slower rise. The results summarized in Table 6.2 indicate, however, that there are differences in the rate of rise estimated by different studies for similar periods (Bard et al., 1989; 1990; Fairbanks, 1989).

There is some evidence that periods of rapid rise in sea level may be due to changes in ice-sheet morphology as melting proceeds (Shaw, 1989; Tooley, 1989). Shaw suggests that drumlins in Saskatchewan, Canada were formed at the end of the last glaciation by catastrophic floods following the

Table 6.2 Sea-level change during the last deglaciation estimated from North Atlantic cores and Barbados coral reefs.

Study	Date (ka BP)	Sea-level change
Bard et al., 1989 North Atlantic	14.5–13.5	rise of 40 m, at a rate of 4 m per 100 a
	~ 12.2	− 67 ± 7 m*/− 83± 10 m**†
	~ 8.5	− 13 ± 11 m**†
	~ 8.2	− 24 ± 8 m*†
Fairbanks, 1989 Barbados	17.1–12.5	rise of 20 m, at a rate of 0.4 m per 100 a
	12	Meltwater Pulse 1A: rise of 24 m in < 1 ka, i.e. 2.4 m per 100 a
	~ 11.0–10.5	minimum rate of rise
	~ 9.5	Meltwater Peak IB: rise of 28 m
Bard et al., 1990 Barbados	13.5	Meltwater Peak
	11.0	Meltwater Peak

* Core SU81–18
** Core SU81–14
† Departure from present-day level

sudden release of subglacially stored meltwater. He estimates that the volume of water was sufficiently large to raise global sea level by about 23 cm in a few weeks (Shaw, 1989). The origin of these drumlins remains open to question and no evidence of flooding has, as yet, been found lower down the valley. Nevertheless, the rapidity and potential impacts of an event of this magnitude justify further investigation (Tooley, 1989). It has also been postulated that the high-frequency (between 10 a and 1 ka) low-amplitude (less than 10 m) oscillations seen in some post-glacial sea-level records may be related to ice-sheet decoupling from the sea floor, leading to rapid collapse before the ice sheet regrounds at a new point (Anderson and Thomas, 1991).

Regional
So far we have been concerned only with global sea-level changes. Regional post-glacial sea-level records are affected by three major factors.

First, there is the glacio-isostatic effect of ice-sheet loading and unloading. The most reliable estimates of the rates of post-glacial uplift are those associated with the Fennoscandinavian ice sheet (Kukal, 1990). The maximum relative uplift is centred near the Gulf of Bothnia and is estimated to be 10 mm a^{-1}. A second, smaller maximum is observed over Scotland. The best estimates for the rate of ongoing isostatic uplift in central northern Scotland are in the range 3–6 mm a^{-1} (Clayton, 1991). It is uncertain whether post-glacial uplift in other regions of the British Isles is complete. A further complication concerns the existence of forebulge regions on the periphery of areas subject to ice-sheet loading. An irregular region of sinking (up to 7 mm a^{-1}) in southern England and northern France, for example, is believed to indicate a former forebulge region (Emery and Aubrey, 1985; Pitty *et al.*, 1992).

Second, there is the effect of water loading along the coast as the present continental shelf was flooded. In the British Isles, it is estimated that this effect operated over the period 18–5 ka BP and caused a maximum depression of about 45 m. This is considerably less than the maximum depression of up to 600 m which is likely to have occurred under the thickest British ice sheets. Estimates of the rate of depression due to water loading around the British coast are given by Clayton (1991). These range from 0.02 mm a^{-1} at Tilbury (on the Thames estuary) to 3.8 mm a^{-1} at Land's End (the tip of the English southwest peninsula).

Third, there is the effect of neotectonic movements. The British Isles is currently subject to downwarping in the southeast and uplift in the west (Pitty *et al.*, 1992). It is, however, difficult to extract the rates of movement.

In order to determine the relative influence of the first two factors on the sea-level history of a particular site, we need to know its location in relation to glacial ice-sheet boundaries. The sensitivity of post-glacial sea-level history to site location is illustrated by a recent study of Holocene glacial rebound and sea-level change in northwest Europe (Lambeck *et al.*, 1990). Three types of site are defined in this study: first, sites such as Angerman River, Gulf of Bothnia, which were located at the centre of the Fennoscandinavian/Scottish ice sheets; second, sites such as Edinburgh located just within the ice-sheet boundaries; third, sites such as Zuid Holland and Biarritz located outside the ice-sheet boundaries. The contribution of three terms to total sea-level change at each site is estimated. These terms are a rigid body term representing the effects of sea-level change on a rigid Earth, an ice unloading term and a water loading term. The model-based calculations show that the ice unloading term dominates at Angerman River and Edinburgh, thus sea level has fallen over the last 20 ka. At Biarritz and Zuid Holland, sea level has risen over the same period.

Additional factors influencing regional post-glacial sea-level trends include local water loading and unloading effects associated with the growth and decay of closed lakes, and the erosion and deposition of sediments (Clayton, 1991). The post-glacial record of sea level from a region such as northwest England reflects the balance of all these various influences (Table 6.3; Tooley, 1974).

6.3.3 Sea-level change over the last century

Estimates of sea-level change over the last century can be made from the global network of tide-recording stations. The world's longest continuous record begins in 1774 and comes from Stockholm in Sweden (Ekman, 1988). Data from tide-gauge stations have to be corrected for the local effects described in the previous section, which may cause the gauge base level to rise or sink. Difficulties in estimating correction factors and the use of different station networks have resulted in varying estimates of mean 'global' sea-level trends.

Table 6.3 Sea-level changes in northwest England over the last 9 ka: the 12 episodes of change identified by Tooley (1974). The transgressions are numbered from 1, the oldest, to 12, the most recent.

Transgression	Date (BP)	Description of changes
1	9200–8500	Rise from –20 m to –9 m OD, at rate of about 1.5 cm a^{-1}. Then fall of over 1 m over 200 a.
2	8400–7800	Slow rise.
3	7800–7600	Short period of very rapid rise, from –9 m to –2 m OD.
4	7600–7200	Slow rise, then fall.
5	6885–6025	Period of oscillating changes.
6	6000–5470	Rising to about + 2 m OD at 5775 a BP.
7	5470–4800	Fall, followed by rapid rise to over +3 m OD, i.e. extensive transgression. Regression over final 500 a.
8	4800–3900	Rise from + 2 m to +3 m OD.
9	3900–3150	
10	3150–2000	Rising, reaching +4 m OD just before 2270 a BP.
11	2000–1300	Maximum of +5.4 m OD reached about 1600 a BP, falling to +4.6 m OD by 1370 a BP.
12	1300–0	No direct evidence of transgression. Maximum of about +5.5 m OD suggested about 800 a BP. Since then has fallen to +3.8 m OD (the present day difference between mean high water and mean sea-level)

The rates and causes of global sea-level change over the last 100 a have recently been the subject of a detailed assessment by the International Panel on Climate Change (IPCC) (Warrick and Oerlemans, 1990). Various estimates of sea-level change are compared, as illustrated in Table 6.4. All the estimates lie in the range 0.5–3 mm a^{-1}. Most lie in the range 1–2 mm a^{-1}. Even among the most recent and presumably, therefore, the most reliable studies, estimates show considerable variation (Gornitz and Lebedeff, 1987; Barnett, 1988; Peltier and Tushingham, 1989; 1991; Trupin and Wahr, 1990). Warrick and Oerlemans conclude that there is no firm evidence in support of an acceleration in sea-level rise over the present century.

Table 6.4 Estimates of global sea-level change over the last 100 a.

Rate (mm a^{-1})	Comments	References
> 0.5	cryologic estimate	Thorarinsson (1940)
1.1 ± 0.8	many stations, 1807–1939	Gutenberg (1941)
1.2 – 1.4	combined methods	Kuenen (1950)
1.1 ± 0.4	six stations, 1807–1943	Lisitzin (1974)
1.2	selected stations, 1900–1950	Fairbridge and Krebs (1962)
3.0	many stations, 1935–1975	Emery (1980)
1.2	many stations/regions, 1880–1980	Gornitz et al. (1982)
1.5	many stations, 1900–1975	Klige (1982)
1.5 ± 0.15	selected stations, 1903–1969	Barnett (1983)
1.4 ± 0.14	many stations/regions, 1881–1980	Barnett (1984)
2.3 ± 0.23	many stations/regions, 1930–1980	Barnett (1984)
1.2 ± 0.3	130 stations, 1880–1982	Gornitz and Lebedeff (1987)
1.0 ± 0.1	130 stations > 11 regions, 1880–1982	Gornitz and Lebedeff (1987)
1.15	155 stations, 1880–1986	Barnett (1988)
2.4 ± 0.9	40 stations, 1920–1970	Peltier and Tushingham (1989; 1991)
1.7 ± 0.13	84 stations, 1900–1980	Trupin and Wahr (1990)

Source: Warrick and Oerlemans, 1990; updated from Barnett, 1985, and Robin, 1986.

The possible sources of error and systematic bias in these data sets have been assessed by the IPCC expert group. First, all studies are based on essentially the same global data set, which contains relatively few long-term tide-gauge records. Second, the geographical distribution of the available records is biased towards northern Europe, North America and Japan. Third, there are problems in removing the effects of local land movements from the individual records.

The IPCC expert group has also assessed the relative contribution of different factors to sea-level change (Warrick and Oerlemans, 1990). The analysis is based on the assumption that the observed 100-year trend is due in part to the 0.5°C increase in global surface air temperature seen over the same period (Jones *et al.*, 1986). Four major climate-related factors with the potential to contribute to sea-level rise are identified: thermal expansion of the oceans; melting of glaciers and small ice caps; melting of the Greenland ice sheet; and melting of the Antarctic ice sheet. The IPCC estimates of the relative contribution of each factor are shown in Table 6.5.

Table 6.5 Estimated contributions to sea-level rise over the last 100 a (cm).

	Low	Best estimate	High
Thermal expansion	2.0	4.0	6.0
Glaciers/small ice caps	1.5	4.0	7.0
Greenland ice sheet	1.0	2.5	4.0
Antarctic ice sheet	−5.0	0.0	5.0
Total	−0.5	10.5	22.0
Observed	10.0	15.0	20.0

Source: Warrick and Oerlemans, 1990.

The thermal expansion estimates are obtained from an upwelling diffusion climate model developed by Wigley and Raper (1987; 1992). Thermal expansion is largely controlled by three model parameters: diffusivity; the ratio of the change in temperature of sinking water to the change in global mean temperature; and climate sensitivity (i.e. the global mean equilibrium $2 \times CO_2 - 1 \times CO_2$ temperature change). The model is driven by past changes in radiative forcing due to increasing anthropogenic greenhouse gas emissions. The first two parameter values are constrained to be consistent with observed global warming, estimated as 0.3–0.6°C over the period 1880–1985. It is estimated that thermal expansion has contributed 2–6 cm to sea-level rise over this period (Wigley and Raper, 1987; 1992).

There is substantial evidence to suggest that a global retreat of glaciers and small ice caps has occurred over the present century (Grove, 1988). The contribution of glacier/ice cap melt to sea level is estimated by assuming that a warming of 1°C will result in a sea-level rise of 1.2 ± 0.6 mm a^{-1} (Meier, 1984; Greuell, 1989; Kuhn, 1992; Oerlemans, 1992). The contribution of the Greenland and Antarctic ice sheets is considered separately because, although it is not known how close either ice sheet is to equilibrium, notable differences in their mass-balance characteristics are found.

Evidence from the Greenland ice sheet indicates strong retreat of outlet glaciers, suggesting that ablation has increased over the last 100 a (Weidick, 1984; Warrick and Oerlemans, 1990). It is considered that this increase is largely due to more melting, runoff and evaporation. It is possible, however, that melting may be partially offset by increased moisture availability and hence increased snowfall. Given this possibility, together with factors such as its greater size, the contribution of the Greenland ice sheet is estimated to be a sea-level rise of 0.3 ± 0.2 mm a^{-1} per 1°C warming. This is considerably less than that of glaciers and small ice sheets.

Finally, the IPCC expert group considered the contribution of the Antarctic ice sheet to sea-level rise over the last 100 a. This assessment was hampered by the lack of data and the likelihood that this ice sheet has not yet adjusted to the last glacial–interglacial transition. Although increased melting, runoff and evaporation are considered to have had a negligible effect over the last 100 a, it is not known whether the Antarctic ice sheet is in balance, or whether it has contributed to the observed sea-level rise (Warrick and Oerlemans, 1990). This uncertainty is reflected in the estimates given in Table 6.5, which range from −5 cm to +5 cm, with a best estimate of 0 cm.

Other factors, such as changes in surface and groundwater storage, may affect sea level on this 100-year time scale. Their possible contributions to sea-level change cannot, however, be estimated on the basis of the few existing data and are not taken into account in the Warrick and Oerlemans study.

Table 6.4 shows the range of uncertainty associated with estimates of the actual rate of sea-level change over the last 100 a. We can compare this with the estimated contributions given in Table 6.5, which lie in the range −0.5 cm to +22 cm. If we take the best estimates from Table 6.5, it appears that thermal expansion, together with the melting of glaciers, small ice caps and the Greenland ice sheet, have contributed 10.5 cm to sea-level rise over the last 100 a. This is just within the range of observed sea-level change indicated by Table 6.5 (10–20 cm). If we assume that the observed values (i.e. 10–20 cm) are reliable, it appears that the best estimates given in Table 6.5 tend to understate sea-level rise. This apparent discrepancy could be rectified by assuming that the contribution of the Antarctic ice sheet is positive, rather than zero or negative.

The uncertainties associated with estimating past sea-level change have considerable implications for our ability to predict future sea-level trends, such as those expected to accompany the enhanced greenhouse effect.

6.4 Greenhouse gas-induced warming and future sea-level change

6.4.1 Introduction

Given the difficulties experienced in explaining the rate of global sea-level rise observed over the last century, it is not surprising that predicting future sea-level change is beset with uncertainty. In Chapter 4, the problems of identifying future greenhouse gas-emission scenarios and of predicting the associated rates and patterns of warming are described. These uncertainties also have an effect on our ability to predict future sea-level rise, which is discussed below. Nevertheless, we demonstrate that estimates of future sea-level change have been refined (Kerr, 1989c; Meier, 1990), indicating that the rate of increase will be slower than earlier studies suggest.

6.4.2 Estimates of future sea-level rise

The IPCC expert group has reviewed estimates of future sea-level change, including some of the most recent studies (Table 6.6; Warrick and Oerlemans, 1990). There are problems in making a comparison because these studies are based on different time periods and on different

Table 6.6 Estimates of future global sea-level rise (cm).

	Contributing factors				Total rise[a]		
	Thermal expansion	Alpine	Greenland	Antarctic	Best estimate	Range[f]	To year
Gornitz et al. (1982)	20	20			40		2050
Revelle (1983)	30	12	13		72[b]		2080
Hoffman et al. (1983)	28–115	28–230				56–345	2100
Polar Research Board (1985)	–[c]	10–30	10–30	−10–100		10–160	2100
Hoffman et al. (1986)	28–83	12–37	6–27	12–220		58–367	2100
Robin (1986)[d]	30–60[d]	20 ± 12[d]	≤+ 10[d]	≤− 10[d]	80[i]	25–165[i]	2080
Thomas (1986)	28–83	14–35	9–45	13–80	100	60–230	2100
Jäger (1988)[d]: results presented at the 1987 Villach workshop					30	−2–51	2025
Raper et al. (1992)	4–18	2–19	1–4	−2–3	21[g]	5–44[g]	2030
Oerlemans (1989)					20	0–40	2025
Van der Veen (1988)[h]	8–16	10–25	0–10	−5–0		28–66	2085

[a] From the 1980s.
[b] Total includes additional 17 cm for trend extrapolation.
[c] Not considered.
[d] For global warming of $3.5°C$.
[f] Extreme ranges, not always directly comparable.
[g] Internally consistent synthesis of components.
[h] For a global warming of 2–4°C.
[i] Estimated from global sea-level and temperature change from 1880 to 1980 and global warming of $3.5±2.0°C$ for 1980–2080.

Source: Warrick and Oerlemans (1990), adapted from Raper *et al.*, 1992.

assumptions concerning greenhouse gas emissions and other factors. Generally, however, sea-level rise is projected to be in the range 10–30 cm by about 2030.

A major uncertainty in the estimation of future sea-level change concerns the ratio of the contributions of thermal expansion and glacier melt. Of the earlier studies listed in Table 6.6, some de-emphasize the role of thermal expansion (Polar Research Board, 1985), whereas others assume the ratio is roughly 1:1 (Gornitz *et al.*, 1982) or give thermal expansion a dominant role (Robin, 1986). Gornitz *et al.* (1982) assume a future thermal expansion to alpine glacier melt ratio of 1:1, although their study ascribes the major part of sea-level rise observed over the last century to thermal expansion. Revelle (1983) estimates that thermal expansion would contribute about 30 cm of the 72 cm rise associated with a global warming of 3–4°C by 2085. Only 25 cm of the total rise is attributed to melting of the Greenland ice sheet and mountain glaciers. An additional rise of 17 cm, based on trend extrapolation, was included. Revelle assigned an uncertainty of ±25% to the total rise of 72 cm. Both the Gorntiz *et al.* (1982) and Revelle (1983) studies assume that the contribution from Antarctica is zero. The Hoffman *et al.* (1983) estimates were derived from model studies which project both global warming and thermal expansion rates. It was assumed that the glacial contribution would be between one and two times the thermal expansion contribution. Estimates were made for four scenarios (low to high) which made different assumptions about the equilibrium $2\times CO_2$ temperature change. The middle scenarios were considered most realistic and gave a range of 144–216 cm by 2100.

The first study to consider in detail the potential contribution of glaciers to sea-level rise was the work of the Polar Research Board (1985). The estimated contribution of alpine glacier and Greenland ice-sheet melt was put at 10–30 cm, with a global warming of 1.5–4.5°C (cf. Meier, 1984). Greater uncertainty surrounded the Antarctic contribution, which was eventually estimated at between −10 cm and +100 cm. Sea-level fall was considered a possibility since more moisture might be carried polewards in a warmer world, to accumulate as snow and ice at high latitudes. The most likely contribution of Antarctic melt was estimated at a few decimetres and a total sea-level rise of 10–160 cm was preferred. As a result of these more detailed assessments of cryosphere melt, Hoffman et al. (1986) repeated their earlier analysis. The revised estimates for thermal expansion and glacier melt were somewhat lower than in their original study. The additional contributions from the Greenland and Antarctic ice sheets, however, mean that the estimates for 2100 are almost identical in both the earlier and the later studies. The Robin (1986) study was based on analysis of the Gornitz et al. (1982) results: linear correlations between the past record of sea level and mean global temperature were calculated. These correlation functions were then applied to the predicted future temperature record. Assuming a global warming of 3.6±2°C over the next century and a thermal expansion:glacier melt ratio of 3:1, a total rise of 25–165 cm by 2080 was predicted.

The range of values of projected sea-level rise in Table 6.6 is large, but can be narrowed on the basis of 'best estimates'. A best estimate for sea-level change by 2050, made on the basis of evidence from the first seven studies listed in Table 6.6, is 30–70 cm (Warrick, 1986). This evidence also indicates a best estimate of 50–200 cm for sea-level rise by 2100. The last four, more recent, studies in Table 6.6 do not provide estimates as far into the future. It is, therefore, difficult to make a direct comparison of the two sets of studies. The more recent work suggests a best estimate of the order of 0–40 cm for sea-level rise by about 2025.

The IPCC expert group has produced its own estimates of future sea-level rise based on the most reliable data and models currently available (Warrick and Oerlemans, 1990). Estimates are made for the four emissions scenarios described in Section 4.2. Discussion here is restricted to estimates of sea-level rise for the Business-as-Usual (BaU) scenario (Tables 6.7 and 6.8). The other scenarios involve successively more severe reductions in anthropogenic greenhouse gas emissions, hence the associated sea-level rise is also reduced.

The estimates for thermal expansion given in Table 6.7 are taken from the upwelling diffusion model developed by Wigley and Raper (1987). The low, best and high estimates are obtained using a climate sensitivity to a doubling of CO_2 of 1.5°C, 2.5°C and 4.5°C, respectively (see Section 4.4.3 for a

Table 6.7 Best estimates of sea-level rise (cm) from 1985 to 2030 assuming the IPCC Business-as-Usual emissions scenario.

	Thermal expansion	Mountain glaciers	Greenland	Antarctica	Total
High	14.9	10.3	3.7	0.0	28.9
Best estimate	10.1	7.0	1.8	−0.6	18.3
Low	6.8	2.3	0.5	−0.8	8.7

Source: Warrick and Oerlemans, 1990.

discussion of climate sensitivity) Additional results from this model are presented by Raper and Wigley (1991).

The estimated contributions of mountain glaciers (Table 6.7) come from a simple global glacier melt model (Raper *et al.*, 1992). This model requires the specification of three global parameters: initial ice volume in 1861 (when the world's glaciers were last estimated to have been in equilibrium), glacier response time, and glacier sensitivity to temperature. The parameter values are chosen to fit the estimated retreat of glaciers over the last 100 a (Grove, 1988).

Table 6.8 Estimated sea-level rise (cm) from 1990 to 2100 for the IPCC Business-as-Usual emissions scenario.

Year	Low	Best estimate	High
2030	8	18	29
2070	21	44	71
2100	31	66	110

Source: Warrick and Oerlemans, 1990.

The estimates for the Greenland and Antarctic ice sheets are based only on observations of the changes in mass balance over the last 100 a and do not account for possible changes in dynamic behaviour such as calving and surging. It is considered that these latter effects operate over time scales greater than 100 a. It is assumed that the Greenland ice sheet is reducing in size and thus contributes positively to sea-level rise. Melting is, however, partially offset by increased snowfall over the higher parts of the ice sheet due to the enhanced hydrological cycle. This latter effect is considered to dominate in the case of the Antarctic ice sheet, hence its contribution to sea-level rise is negative.

The estimates given in Table 6.7 are obtained by applying temperature sensitivity values of $+0.3\pm0.2$ mm a^{-1} per 1°C warming for the Greenland ice sheet and -0.3 ± 0.3 mm a^{-1} per 1°C warming for the Antarctic ice sheet. The large error bars reflect the uncertainty in estimating the relative balance of conflicting processes (for example, increased ablation versus increased accumulation). Estimates of temperature change are taken from transient runs of fully coupled atmosphere–ocean GCMs (Stouffer *et al.*, 1989). Based on these model results, it is assumed that warming over Antarctica will be the same as the global mean warming, while warming over Greenland will be enhanced by a factor of 1.5.

For the BaU scenario, it is estimated that global mean sea level will rise by 8–29 cm by 2030, with a best estimate of 18 cm (Table 6.7). This range encompasses the majority of estimates given in Table 6.6. For 2100 the rise estimated by the IPCC is 31–110 cm, with a best estimate of 66 cm (Table 6.8). The range of uncertainty is large and increases with time. It is, however, apparent that thermal expansion and the retreat of mountain glaciers will continue to provide the largest contribution to sea-level rise over the next 100 a. It is also reasonable to conclude that a rise of more than 1 m over the next 100 a is unlikely (Warrick and Oerlemans, 1990). If effective controls of anthropogenic greenhouse gas emissions are introduced then the rise may be reduced to about one-third of that associated with the BaU scenario. However, even under the strictest emission controls, sea level will continue to rise. The influence of lags, introduced by the thermal inertia of the oceans and the characteristic response time of ice sheets to climate, means that the sea-level rise 'commitment' is greater than the actual rise at any given time (Warrick and Oerlemans, 1990; Raper and Wigley, 1991).

Beyond 2100, the greatest uncertainty in estimates of sea-level rise relates to the possible rapid disintegration of the West Antarctic ice sheet if warming continues.

6.4.3 The Antarctic ice sheet

The sensitivity of the Antarctic ice sheet to warming is uncertain. The fact that it survived the warmth of the Holocene thermal optimum (about 6 ka BP), when global temperature was about 1°C greater than today, implies that it is not close to collapse at the present time. Whether or not it survived the previous interglacial (about 125 ka BP) is, however, controversial. It has been suggested that it did, in fact, collapse during this earlier episode, contributing about 5 m to global sea level (Mercer, 1978). The alternative view is that most of the estimated 6 m increase in sea level at the last interglacial can be attributed to the partial, or total, collapse of the Greenland ice sheet (Koerner, 1989).

Two principal mechanisms which could lead to the rapid disintegration of the Antarctic ice sheet can be identified (Van der Veen, 1987). The first has been described as the 'fringing ice-shelves' hypothesis (Mercer, 1978; Hughes, 1987; Van der Veen, 1987; 1988).

The West Antarctic ice sheet is surrounded by large ice shelves which control its stability. If the grounding line of these ice shelves begins to retreat because of rising sea level and/or thinning of the ice shelves, then the retreat may become irreversible, eventually leading to the collapse of the entire ice sheet (Van der Veen, 1987). Thinning of the ice shelves could occur due to the effect of ice streams drawing ice out from the main ice sheet (Hughes, 1987; Van der Veen, 1987). The fastest-flowing ice stream known today is Jakobshavns Glacier in Greenland, which has an average velocity of 7 km a^{-1}. Typical outflow velocities for Antarctic ice streams are 0.5 km a^{-1}. If, however, the discharge velocity of Antarctic ice streams were to approach that of Jakobshavns Glacier, it would take at least 300–400 a to destroy the entire West Antarctic ice sheet (Van der Veen, 1987). This scenario, which should be viewed as a 'disaster' scenario, would cause a 6 m rise in global sea level.

The response of the grounding line of the West Antarctic ice sheet to a warming climate has also been assessed in modelling studies (Lingle, 1985; Budd et al., 1987). In Lingle's (1985) model, the grounding line does not start to retreat until the time at which the ocean water under the Ross Ice Shelf has warmed sufficiently to enhance melting (estimated to be 2085). Lingle's worst-case scenario indicates that the West Antarctic ice sheet will take at least 600–700 a to collapse. A number of the assumptions made in this worst-case assessment are considered to be unrealistic (Van der Veen, 1987; 1988).

The second mechanism which may affect the Antarctic ice sheet is surging (Van der Veen, 1987). It is not considered likely that a single surge event could affect the entire Antarctic ice sheet because it is made up of a number of individual dynamic glacial systems, each behaving differently (Van der Veen, 1987). If surges do occur, they are more likely in East Antarctica, where relatively fast ice velocities are already observed, than in West Antarctica.

Budd et al. (1987) have used a hierarchy of models to investigate the future behaviour of the Antarctic ice sheet. They conclude that the total contribution of this ice sheet to global sea-level rise may be up to 1 m after 500 a, and up to 3.5 m after 1 ka. The best evidence currently available suggests that the collapse of the West Antarctic ice sheet could contribute, at maximum, another 5 m to global sea-level rise. This would not occur for at least 500 a. Given the number of conditions that must be met (Hughes, 1987; Van der Veen, 1988), this scenario should be ascribed a low, but not zero, probability.

6.5 Conclusions

In the immediate future, the rise in global sea level observed over the last 100 a is projected to continue. Indeed, the rate of increase is expected to accelerate due to the enhanced greenhouse effect. We consider that the best estimates of the rate of rise over the last 100 a lie in the range 1–2 mm a^{-1}. The best estimates for the next 100 a lie in the range 2–10 mm a^{-1}. There is also a remote, but non-zero, possibility that an additional sea-level rise of 5 m may occur sometime during the next 500–1000 a due to the collapse of the West Antarctic ice sheet. Chapter 5 discussed the uncertainties connected with the enhanced greenhouse effect, in particular with respect to future emissions scenarios and the longer-term persistence of the effect. These uncertainties mean that it is not possible to estimate the total sea-level rise which may occur in the future due to enhanced greenhouse warming.

One of the furthest projections into the future that has been made is the model-based study of the Greenland ice sheet carried out by Huybrechts *et al.* (1991). This study suggests that the Greenland ice sheet may contribute up to 30 cm to global sea-level rise by 500 a after the onset of warming (i.e. by AD 2350). In comparison, the melting of the entire Fennoscandinavian ice sheet during the last deglaciation is estimated to have contributed 15 m to global sea-level rise. We conclude that the maximum greenhouse gas-induced sea-level rise is likely to be in the order of decimetres, rather than metres or tens of metres.

On longer time scales, global sea-level change is dominated by large glacial–interglacial fluctuations. A constraint on future fluctuations of +6 m for interglacial episodes and −130 m for glacial periods is provided by the record of the previous interglacial–glacial cycle.

The estimates discussed above all relate to global sea level. In order to estimate regional sea level it is necessary to account for a range of local factors, many of which involve geological processes such as tectonic and isostatic movements.

The social and economic impacts of rising sea level during a period of enhanced greenhouse warming are potentially substantial. The possibility of direct inundation due to continuing sea-level rise is clearly of concern to low-lying nations such as Bangladesh and the Netherlands, and to cities such as Venice and London. It is likely that higher-lying areas may also feel the influence of even a moderate rise in sea level. Such regions may be at increased risk from flooding due to storm surges, salt intrusion, increased coastal erosion and changes in the rate of river incision (Gornitz and White, 1991).

In the longer-term future, sea level can be expected to fall with the onset of the next glaciation. In present-day coastal areas, falling sea level is initially likely to be accompanied by a reduction in the frequency and strength of extreme events such as storm surges. The major climatic effects of the very large falls in sea level during glacial episodes relate to the associated increase in continentality at specific locations.

7 Regional sequences of past climate

7.1 Introduction

In this chapter, we assess regional sequences of Quaternary climate change. As outlined in Chapter 1, a main aim of this book is to understand how regional patterns of climate change relate to larger-scale and long-term forcing mechanisms. To illustrate these relationships, we take as our example one particular region of western Europe, the British Isles. The general background pattern of climate change over the Quaternary period in the British Isles is described. Chapter 8 provides a more detailed picture of the climatic conditions experienced in this study region over the last glacial–interglacial cycle.

In order to reflect the range of climate variability and data availability within a region such as the British Isles, and to address the needs of climate impact studies in terms of spatial resolution, we consider three areas in detail. Their location is shown in Figure 7.1. Places referred to in the text are also shown in this figure.

The first two, Caithness in northeast Scotland and Cumbria in northwest England, offer a useful contrast between an area which has been relatively well studied (Cumbria) and one whose glacial and climatic history is less well known (Caithness). They also have contrasting topography. Cumbria is clearly divided into the mountainous Lake District, with a maximum altitude of about 1000 m, and the low-lying coastal plains. In Caithness, the land slopes more gently towards the sea from an inland plateau, which has a typical altitude of under 300 m. There are differences in coastal topography. The gentle contours and slopes of the Irish Sea off the Cumbrian coast contrast with the steeper gradients off the northeast coast of Scotland. Caithness and Cumbria have been investigated by the Climatic Research Unit in a project to assess future climate states in the UK over very long time scales, for input to studies of the post-closure performance of radioactive waste repositories (Goodess *et al.*, 1992). (At the time of writing no such studies relating to Caithness are being undertaken. Sellafield in Cumbria is, however, being studied as the possible location of an underground repository.)

In order to extend the geographical distribution of our study areas, we also consider Wales. This third area has the same advantage as Cumbria, in that a large body of literature exists on its palaeoclimate history. It is a predominantly mountainous area, within which two major ranges can be identified: Snowdonia in the north, with a maximum altitude of over 1000 m; and the Brecon Beacons in the south.

Wherever possible throughout this chapter we have elected to discuss events as they occurred, from the oldest to the youngest. Although this policy has been followed almost universally in the text, it has not always

Figure 7.1 Location of the British Isles study areas and places discussed in the text.

appeared logical in the arrangement of the tables. This is because there is a lack of consistency in the numbering schemes: on the one hand, oxygen isotope stages are numbered from 1, the youngest, whereas pollen-based series are numbered from 1, the oldest. The caption of each table indicates the ordering of events.

A few relatively short post-glacial pollen cores are available from Caithness (Keatinge and Dickson, 1979; Peglar, 1979; Walker, 1984) and the Cumbrian lowlands (Pennington, 1965; Walker, 1966), and from Wales (Hibbert and Switsur, 1976; Lowe, 1981; Stevenson and Moore, 1982; Walker, 1982; Heyworth et al., 1985; Walker and Harkness, 1990). No long continuous reconstructions of past climate are available. The sequence of past climate in our three study areas must, therefore, be reconstructed using evidence from elsewhere in the British Isles and northwest Europe.

First, there is land-based evidence from the British Isles, taken from pollen and beetle fossil records, faunal assemblages, speleothem growth, glacial till and ice wedges. Geomorphological and sedimentary features provide fragmentary evidence of past glacial episodes (Bowen et al., 1986). Evidence of individual warm interglacial and interstadial periods comes from pollen records and from fossil faunal assemblages. Chronologies of these warm periods have been reconstructed from speleothem records (Stringer et al., 1986; Gordon et al., 1989) Unfortunately, most of the land-based records provide only a qualitative description of climate conditions. Records which provide estimates of, for example, mean annual or summer temperature are rare.

Second, there are long continuous records from sediment cores taken from the surrounding oceans. Records are available from the North Sea and from the North Atlantic (Ruddiman and McIntyre, 1976; Duplessy et al., 1986; Ruddiman et al., 1986; Boyle and Keigwin, 1987; Jensen and Knudsen, 1988; Shackleton et al., 1988; Overpeck et al., 1989). Of these, the climate records from the North Atlantic provide a more valuable indicator of past climate than do records from the North Sea. The climate of the British Isles is strongly influenced by sea-surface conditions in the Northeast Atlantic (Perry and Walker, 1977; Flohn and Fantechi, 1984; Lockwood, 1985). Research has shown that climate change over time scales of hundreds to thousands of years is similar, and synchronous, in Iceland and the British Isles (Lamb and Johnson, 1966; Lamb, 1977a). Reasoning alone suggests that, in the mid-latitude westerly bands, westernmost islands must be most strongly influenced by circulation systems approaching from the Northeast Atlantic.

Third, long pollen records from the European continent (Guiot et al., 1989; de Beaulieu et al., 1991) may be used as a guide to conditions in the British Isles. These continuous records can also be used to interpret evidence from the first two data sources and, in particular, to explore relationships between the British land-based and oceanic evidence.

Each of these three sources of evidence is discussed in detail below. We then proceed to examine correlations between the land-based and ocean-based records, and between regional- and global-scale records. This is a relatively easy task for the last glacial–interglacial cycle (the last 125 ka), for which many high-resolution records are available, but is difficult further back in time. Finally, we show that similar correlations can be established for other regions. Examples are given from North America and Africa.

7.2 Land-based records from the British Isles

7.2.1 *Cold episodes*

Chronologies of glacial events in the British Isles are based on stratigraphic and geomorphological evidence, including pollen records and the presence of physical features such as glacial till and erratics. While it is relatively easy to find evidence of glaciation, it is much more difficult to identify and date individual events so that reliable chronologies can be constructed. Early chronologies, such as the widely used standard of Mitchell *et al.* (1973), have subsequently been revised. A number of uncertainties remain concerning the number, timing and extent of individual glacial events, but a more complex series of multiple Quaternary glaciations is now preferred.

The available evidence of Quaternary glaciation in the Northern Hemisphere has been extensively reviewed (Sibrava, 1986). One section of this review assesses the evidence for Quaternary glaciations in England, Scotland, Ireland and Wales and establishes a provisional correlation between individual, regional, events within the British Isles and dated climate stages identified in ocean core records (Bowen *et al.*, 1986). The resulting chronologies are shown in Figure 7.2. The geographical extent of the major glacial episodes is shown in Figure 7.3. It should be noted that this figure represents only one interpretation of the available evidence. Other authors have proposed somewhat different glacial boundaries.

Glaciation in Scotland

The clearest evidence for recurrent glaciation in the British Isles comes from Scotland. The evidence assessed by Bowen *et al.* (1986) supports the occurrence of at least four episodes of major glaciation in Scotland in the last 700 ka. Three of these events occurred during the Middle Pleistocene and are referred to, from the oldest to the most recent, as the 'Glacial A' of the Cromerian Complex, the Anglian and the Wolstonian. Very few Early or Middle Pleistocene sites are known. Much of the available evidence for these glacial episodes comes from the North Sea Basin. The fourth, and most recent, glaciation is known as the Devensian.

The beginning of the last interglacial (the Ipswichian) is conventionally taken to mark the beginning of the Late Pleistocene. Although Scottish evidence from this interglacial is scarce, evidence for the Devensian glacial period, which followed, is more extensive, particularly for the latter stages. There are some rather tentative indications of an ice advance during the Early Devensian (Jardine *et al.*, 1988). The evidence for ice sheet expansion during the Late Devensian and during the subsequent Loch Lomond Stadial (the Younger Dryas cold period) is convincing. The major features of the

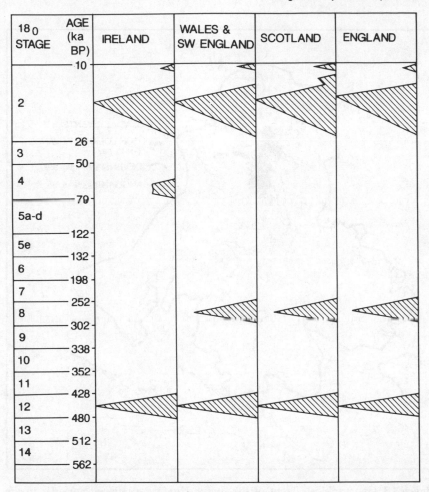

Figure 7.2 Revised chronology of British glacial episodes identified by Bowen *et al.* (1986). The wedge symbols indicate periods of ice advance and retreat. So far as is possible, the physical evidence for glaciation has been fixed in time using four different dating methods (^{14}C, thermoluminescence, amino acid ratios, uranium series). A provisional correlation with the oxygen isotope stages of Shackleton and Opdyke (1973; 1976) is indicated. (After Bowen *et al.*, 1986.)

Late Devensian glaciation in Scotland are summarized below.

Build-up of both the Scottish Mainland and the Scandinavian ice sheets may have begun about 75 ka BP. Early studies imply that the Scottish Mainland ice sheet had one centre and that the whole of Scotland and most of the North Sea Basin were ice-covered during the Late Devensian glaciation (Boulton *et al.*, 1977). However, it is now considered that the Scottish ice sheet was polycentric and that the relative strength of individual ice flows varied over time. The ice centres in northern Scotland were relatively weak. It has been suggested that the Southern Uplands eventually

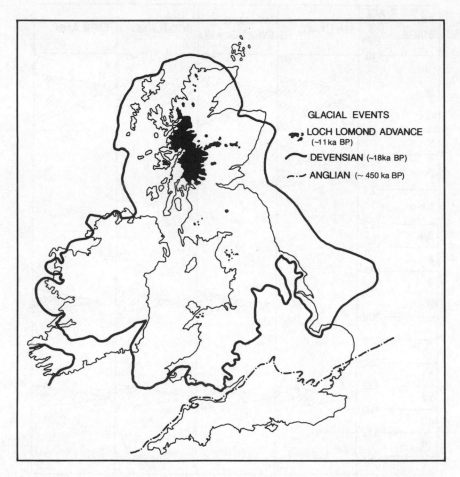

Figure 7.3 Geographical extent of the major Quaternary glacial episodes in the British Isles. (From Bowen *et al.*, 1986.)

succeeded the Highlands as the major centre of Scottish glaciation (Bowen *et al.*, 1986). This may, in part, be explained by a southward shift of the zone of maximum effective precipitation.

It is unlikely that the ice sheet was continuous or of uniform thickness. There is evidence of both low-lying and high-level ice-free regions, known as *enclaves* and *nunataks*, respectively. According to the ice-sheet limits of Bowen *et al.* (1986), parts of Caithness were unglaciated during the Late Devensian (Figure 7.3). Pitty (1992), however, considers that reconstructions showing Caithness and Orkney to be ice-free during the Late Devensian should be treated with caution. Nevertheless, it has been established that the North Sea to the southeast of the Shetland Islands and to the east of the Moray Firth was ice-free at the height of the Late Devensian (Flinn, 1981). Evidence also supports the existence of isolated local ice caps in the Shetland and Outer Hebridean islands.

The movement of the Scottish Mainland ice sheet was influenced by its polycentric nature, and by the presence of other ice sheets. The expansion of the Southern Uplands and Irish ice sheets, for example, blocked the southerly movement of the Scottish Highland ice which was consequently deflected westwards. In the early stages of the Late Devensian glaciation, ice flowed northeastwards into Caithness. As glaciation proceeded, the ice was reoriented to flow northwest, probably because of the encroachment of ice from the Moray Firth (Hall and Whittington, 1989).

It is considered that the Scottish ice sheets began to retreat at about 13.5–13.0 ka BP (Bowen et al., 1986). It is likely that melting was rapid, but it is not certain whether deglaciation was total by the time of the Loch Lomond Stadial. The maximum of this last ice advance in Scotland occurred shortly after 10.9 ka BP and its limits, as estimated by Bowen et al. (1986), are shown in Figure 7.3.

Glaciation in the Lake District and Cumbrian lowlands

The revised sequence of English glaciation proposed by Bowen et al. (1986) is shown in Table 7.1. This sequence is still subject to many uncertainties and only the Anglian and Devensian glaciations can be reconstructed with any degree of confidence.

There is little evidence for glacial episodes in the Lake District prior to the Late Devensian but, as in Scotland, the evidence for Late Devensian glaciation is convincing. This glaciation is sometimes referred to as the Dimlington glaciation. At the height of the Late Devensian glaciation, it is estimated that the Lake District ice cap was at least 760 m thick, covering all but the highest mountain peaks (Walker, 1966). The ice cap extended to the surrounding plains as far south as the West Midlands (Pitty, 1992). In west Cumbria, ice was deflected westwards and southwards down the coast by Scottish ice encroaching from the north on to the Carlisle Plain and coastal lowlands. The Cumbrian ice also met Irish Sea ice advancing from the west.

The maximum limits of the Devensian glaciation were reached at about 18 ka–17 ka BP. The limits proposed by Bowen et al. (1986) are shown in Figure 7.3. Extensive deglaciation had occurred by about 14.5 ka BP, by which time ice had retreated northwards at least as far as the lowlands of the Lake District. This glacial retreat continued until the onset of the Loch Lomond Stadial or Younger Dryas.

In England, evidence for glacial readvance during the Loch Lomond Stadial mainly takes the form of glacial landforms and varved lake sediments. Ice began to accumulate in the highest parts of the Lake District mountains, reoccupying former sites of Late Devensian corrie glaciation, about 11 ka BP (Pennington, 1970; 1978) and remained present for about 450 a. In the areas of glacial advance (the Lake District and the Pennines in England, and Snowdonia in Wales) and in other parts of England, the impacts of this climatic deterioration are shown by accelerated river development, mass movement and cryoturbation (Kerney, 1963; Kerney et al., 1964; Johnson, 1975; Rose et al., 1980). This is the last time that glacier

Table 7.1 English Pleistocene glacial stratigraphy. Revised series developed by Bowen *et al.*, 1986.

	Interglacial stages	Glacial stages	Glacial episodes
Late Pleistocene	Flandrian		
			Loch Lomond glaciation
		late	
		Devensian	Dimlington glaciation (named after the
		middle	type-site on the northeast English
		early	coast)
	Ipswichian		
Mid Pleistocene			Glaciation in NE England
	Hoxnian		
		Anglian	Lowestoft/N Sea Drift glaciation
	Cromerian		
Early Pleistocene		Beestonian	Glaciation in W Midlands and N Wales
	Pastonian		
		Pre-Pastonian	Glaciation in W Midlands and N Wales
			Glaciation in W Midlands and N Wales
			Glaciation in W Midlands and N Wales
		Baventian	Glaciation in N Sea region
		Bramertonian	
	Antian	Thurnian	
	Ludhamian		
		Pre-Ludhamian	

ice was present in England and Wales, albeit confined to the higher northern and western mountains.

Glaciation in Wales
According to the review of Bowen *et al.* (1986), there is no direct evidence, such as glacial till, for Early Pleistocene glaciation in Wales. Such a glaciation has, however, been inferred by tracing glacial erratics from the Berwyn and Snowdonia mountain ranges in north Wales eastwards as far as East Anglia.

The sequence of Welsh glaciation for the Middle and Late Pleistocene proposed by Bowen *et al.* (1986) is shown in Table 7.2. Within the Middle Pleistocene period, there is convincing evidence of an extensive 'Irish Sea' glaciation in Wales, equivalent to the Anglian glaciation in Scotland and England. During this period, Wales was covered by local ice sheets and

Table 7.2 Welsh Middle and Late Pleistocene glacial sequence proposed by Bowen *et al.* (1986)

	Event	Date
Middle Pleistocene	Irish Sea glaciation	equivalent to Anglian glaciation in Scotland and England
	Paviland glaciation (S Wales only)	older than 225 ka BP
Late Pleistocene	Late Devensian glaciation	comparable to event in Scotland and England
	Gwynedd Readvance (N Wales only)	occurred sometime in period 18–14.6 ka BP
	Loch Lomond Advance/Younger Dryas	~10.6–10.0 ka BP

glaciation was also extensive in the southwestern and southern borders, including the Bristol Channel. The southern extent of this glaciation is shown in Figure 7.3 (Bowen *et al.*, 1986). It follows the southern coastline of the Bristol Channel and reaches the Isles of Scilly. The major direction of ice movement over Wales was from the Irish Sea, and from northwest to southeast. Erratics from this flow have been found as far east as Cardiff on the Bristol Channel. There is also some evidence from south Wales in support of a later Middle Pleistocene glaciation, known as the Paviland (Bowen *et al.*, 1986). This possible event is named after the Paviland moraine on the Gower coast of south Wales. The presence of this moraine suggests that an ice mass moved across an area previously glaciated during the Irish Sea event. The local stratigraphy implies that the Paviland event is older than about 225 ka BP. No equivalent glacial deposits have, however, been found elsewhere in the British Isles.

The Late Pleistocene Welsh stratigraphy is constrained by amino acid geochronologies and uranium-series dates from raised beaches in south Wales (Bowen *et al.*, 1985) and by radiocarbon dates on cave fauna (Rowlands, 1971; Stringer *et al.*, 1986). The age and extent of Late Devensian glaciation in Wales is considered to be reliably established (Bowen *et al.*, 1986). The limits are shown in Figure 7.3. The southern extent is estimated from dated shells in the glacial drifts found in the south and southeast coastal areas.

There is some evidence of a glacial readvance, termed the Gwynedd Readvance, in north Wales after the Late Devensian maximum at 18 ka BP (Bowen *et al.*, 1986). Deglaciation was, however, complete in the northwest Lleyn Penninsula no later than 14.46 ka BP (Coope *et al.*, 1971).

During the Loch Lomond Stadial/Younger Dryas event, cirque glaciers in Snowdonia (Lowe, 1981; Gray, 1982) and the Brecon Beacons (Walker, 1980; 1982) readvanced. Gray (1982) has identified 35 former cirque glaciers, covering a total area of 17.5 km^2, which are attributed to the Younger Dryas event. The onset of the Younger Dryas in the Brecon Beacons is estimated at about 10.6 ka BP (Walker, 1980; 1982), which is somewhat younger than dates estimated for north Wales (Lowe, 1981; Heyworth *et al.*, 1985) and other regions of the British Isles. Throughout Wales, however, climate amelioration had set in by about 10 ka BP.

7.2.2 Warm episodes

The English Pleistocene sequence shown in Table 7.1 identifies seven interglacial periods. The most widespread information on individual interglacial events comes from pollen records (Mitchell *et al.*, 1973; Bowen, 1978). Additional evidence comes from fossil Coleoptera (beetle) records (Coope, 1975; Atkinson *et al.*, 1987; Kashiwaya *et al.*, 1991), dated raised beaches (Bowen *et al.*, 1985; 1989; Bowen and Sykes, 1988), and from fossil faunal assemblages (Gascoyne *et al.*, 1981; Stringer *et al.*, 1986). Much of the evidence is restricted to the two most recent interglacials, the Ipswichian and the Holocene (or Flandrian).

Evidence from pollen and fossil Coleoptera records also supports the existence of warm remissions during cold episodes, known as interstadials. Early Coleoptera studies, for example, suggest that three interstadials (the Wretton, Chelford and Upton Warren) occurred, in lowland England at least, during the last glaciation (Coope, 1975). More recent studies suggest that the sequence of events in the Late Pleistocene may have been considerably more complex (Stringer *et al.*, 1986; Gordon *et al.*, 1989).

The occurrence of speleothems, such as stalagmites, in limestone caves is dependent on groundwater recharge and on biogenic carbon dioxide production in the soil. Speleothem growth is, therefore, related to temperature and to water availability. Extensive speleothem growth indicates warm and relatively wet climate conditions, typical of interglacial and interstadial periods.

In the most comprehensive study of Britain to date, Gordon *et al.* (1989) have constructed a composite speleothem growth frequency curve for the last 220 ka. Five hundred and twenty-one speleothem samples were taken from 11 regions of the British Isles, from Sutherland in the north to Jersey in the south, and dated by the uranium-series method. If the frequency distribution of these dates is plotted, the frequency peaks define periods of active speleothem growth. The speleothem growth peaks may in turn be used to define periods of increased vegetation growth and climatic warming in interglacial and interstadial episodes. Three interglacial peaks (Ipswichian 3 to 1) and seven interstadial peaks (the Pre-Ipswichian 1 and Devensian 6 to 1) can be identified over the period 220 ka–20 ka BP and are shown in Figure 7.4.

Correlations have been established between some of the interglacial peaks in the speleothem record and other, independent, reliably dated records from the Ipswichian interglacial. For example, faunal assemblages from Bacon Hole cave in Wales, indicative of interglacial conditions, have

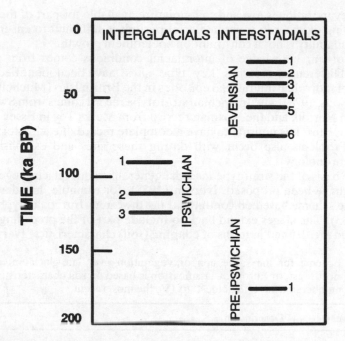

Figure 7.4 Best estimate ages for speleothem growth frequency peaks in the British Isles. (Data from Gordon *et al.*, 1989.)

been dated to 125 ka BP (Stringer *et al.*, 1986). Similar faunal remains from Victoria Cave, Yorkshire, have been dated to 120±6 ka BP (Gascoyne *et al.*, 1981). The Bacon Hole and Victoria Cave assemblages can be linked with the Ipswichian 3 peak in the speleothem record, dated to 124 ka BP. Such correlations are not possible for the interstadial episodes because of the lack of dated records (Gordon *et al.*, 1989).

This composite speleothem curve provides a unique dated chronology of warm episodes in the British Isles over the last interglacial–glacial cycle. Unfortunately, it is not possible to infer anything about the relative or absolute magnitude of temperature or precipitation in the warm episodes identified by this chronology.

Milankovitch frequencies of about 40 ka, 23 ka and 19 ka (see Chapter 2) have been identified in this speleothem record (Kashiwaya *et al.*, 1991). The record has also been compared with the insolation record for caloric summer at 55°N. Agreement is generally good before 100 ka BP and after 70 ka BP, but is less good in the intervening period, i.e. in the period between the Ipswichian interglacial and the Last Glacial Maximum (Kashiwaya *et al.*, 1991). The authors speculate that temperature may be the dominant controlling factor on speleothem growth in glacials, and water availability in interglacials. A combination of these factors is thought to operate in transitional periods. Although precipitation is lower in glacial periods and groundwater may be frozen in the coldest periods, the authors argue that, because evaporation is low and soil moisture available for part of the time, the dripping of water in caves in glacial periods is sufficient to ensure that water availability is not a constraint on speleothem growth.

Much of our knowledge of interglacial conditions comes from pollen records. Different stratotypes (key 'type' sites) have been identified for a number of individual interglacial episodes in the British Isles (Mitchell *et al.*, 1973; Bowen, 1978). The Ipswichian stratotype record comes from Swanton Morley in Norfolk and the Hoxnian record from Marks Tey in Essex. Even at the best sites, it is unusual to have a complete record of each interglacial period. Problems also occur with dating these sites and establishing a reliable chronology.

On the basis of the stratotype records, generalized interglacial vegetation schemes have been proposed. Iverson (1958), for example, has devised a four-stage scheme based on continental northwestern European type areas (Table 7.3). The stages extend back to include part of the preceding Late-glacial and are defined in terms of edaphic (soil) characteristics. Iverson's

Table 7.3 Scheme for the succession of vegetation over late-glacial/interglacial periods in northwestern Europe. Classification is based on soil characteristics. The stages are numbered from I, the oldest, to IV, the most recent.

Stage name	Description
Iverson I (Cryocratic phase)	Cold and dry climate, with active cryoturbation. Sparse assemblages of pioneer, arctic-alpine, steppe and ruderal herbs grow on base-rich, skeletal mineral soils.
Iverson II (Protocratic phase)	Temperature rising. Soils are unleached and fertile, but low in humus. Ample light and space for immigrating flora. Shade-intolerant species form grassland, scrub and open woodland.
Iverson III (Mesocratic phase)	Temperature may continue to rise, but not essential. Shade-intolerant species are replaced and eliminated. Fertile brown soils develop. Temperate deciduous forests are established.
Iverson IV (Telocratic or oligocratic phase)	Retrogressive stage of soil deterioration to infertile humus-rich podsols and peats. Open, conifer-dominated woodlands, ericaceous heaths and bog communities dominate and, during the Holocene, heather. Temperature may fall as soil conditions decline.

Source: Iverson, 1958.

scheme describes an interglacial succession from herbs growing on skeletal mineral soils at the start, to temperate deciduous forests on fertile brown soils in the warmest (mesocratic) phase. This scheme has been adapted by Birks and Peglar (1979) to reflect the observed succession in the Shetland Islands. A second scheme, based on vegetation rather than edaphic characteristics, has been proposed by Turner and West (1968) and is shown

Table 7.4 Scheme for the succession of vegetation over interglacial periods in East Anglia, UK. Classification is based on vegetation, rather than soil, characteristics. The stages are numbered from 1, the oldest, to IV, the most recent.

Stage name	Description
West I (Pre-temperate substage)	Forest vegetation develops, dominated by Boreal trees (birch and pine). Light-demanding herbs and shrubs also occur. Pollen frequencies of elm and oak begin to increase.
West II (Early-temperate substage)	Mixed oak forest develops and expands (elm, oak and hazel). Dense forest develops on fertile soils. Towards end of stage, lime appears.
West III (Late-temperate substage)	At beginning of stage, hornbeam, conifers and beech expand. At same time former mixed oak forest declines. Changes may be due to soil deterioration, but towards end of stage climate changes become important.
West IV (Post-temperate substage)	Temperate forest trees disappear and are replaced by Boreal trees (pine, birch, spruce and alder). Forest thins and open heaths develop. Ericaceous heaths give way to grassland as climate continues to deteriorate.

Source: Turner and West, 1968.

in Table 7.4. This scheme is based on East Anglian type areas only and does not cover the cryocratic phase of Iverson.

Most of the stratotypes come from regions to the south of the ice-sheet limits, notably from southern and central England. Geographical differences in vegetation exist at the present day, and we can assume that they existed in the past. It would be unreasonable to assume, therefore, that the evidence from the stratotypes can be applied to our three study areas. Only small amounts of evidence are available for vegetation changes in the three regions, and these are reviewed below.

Pollen records from the Caithness region
Post-glacial pollen records are available from southeast Caithness and the neighbouring Orkney islands (Keatinge and Dickson, 1979; Peglar, 1979). Walker (1984) has reviewed the late Quaternary pollen records available for Scotland.

Four cores from Glimms Moss on the Orkney mainland have been analysed (Keatinge and Dickson, 1979). The longest is 6 m in length. Two horizons from this latter core have been given ^{14}C dates of 5681±55 a BP and 2145±65 a BP. Four local pollen assemblage zones have been identified and their major characteristics are outlined in Table 7.5.

The earliest levels of the Orkney cores, dating from around 6–5 ka BP, show the presence of birch-hazel scrub with a rich ground flora of tall herbs

Table 7.5 Summary of the four pollen assemblage zones identified in the Glimms Moss cores, Orkney. The zones are numbered from GM1, the oldest, to GM4, the most recent.

Zone	Characteristics of the pollen assemblage zones
GM1	High arboreal pollen content, characterised by woodland (birch and hazel) and scrub. Evidence of elm decline about 5 ka BP, birch/hazel decline about 4.7 ka BP and reduction in wind-transported pine pollen about 4.5–4 ka BP.
GM2	Low arboreal pollen. Gramineae (reeds/wild grasses) dominate but fluctuate. Some evidence of agricultural influences.
GM3	Little information about regional vegetation changes. Arboreal pollen content greater than GM2 but Gramineae also present.
GM4	High calluna and sphagnum content, but reduction in Cyperaceae. As GM3, reveals little about regional vegetation. Likely that GM3 and GM4 saw local resurgence of scrub and more open vegetation.

Source: Keatinge and Dickson, 1979.

and ferns. Oak, pine and alder were not present. The record shows the classic elm decline which was closely associated with the earliest Neolithic settlements on Orkney (about 4.8 ka BP). This transition from scrub woodland to pasture and tall herb vegetation may also be associated with increased wind speeds, as indicated by sand-blown features in the nearby Bay of Skaill. Blanket peat formation on the hills from about 3.4 ka BP onwards may be related to a more oceanic climate and/or to increased grazing pressure (Keatinge and Dickson, 1979).

A much longer pollen record is available from the Loch of Winless core in southeast Caithness (Peglar, 1979). Ten radiocarbon dates have been obtained for this 6 m long core. The core dates from 12 690 a BP and has been subdivided into five local assemblages (Table 7.6). Zones LW1 and LW2 date from the end of the Late Devensian while Zones LW3, LW4 and LW5 come from the Holocene (10 ka BP to present).

The Loch of Winless record indicates a lack of tree cover since the last glaciation. In the climate amelioration at about 10 ka BP the typical Late Devensian dwarf shrub and herb communities were replaced by rich tall herb and fern communities. There is evidence of patchy willow and hazel scrub in sheltered sites. Finally, heather-dominated heaths developed on the well-drained acidified soils of the upper core levels.

The lack of trees growing in southeast Caithness in the post-glacial period indicates the greater similarity of this region to the Orkney and Shetland islands than to neighbouring parts of the mainland such as Sutherland. These islands still lack extensive woodlands today, in part because of factors such as relatively low summer temperatures, extreme wind exposure, low

Table 7.6 Summary of the five local pollen assemblages identified in the Loch of Winless core, southeast Caithness. The zones are numbered from LW1, the oldest, to LW5, the most recent.

Pollen zone and date (a BP)	Characteristics
LW1 ~12 690–12 000	>80% herb and 7–13% tree pollen, indicating open treeless vegetation (dwarf-shrub heathland). Possible marine incursion ~12.2 ka BP.
LW2	Pollen content still very low.
LW2a 12 000 11 250	Mosaic of communities, spread of low-growing dwarf-shrub communities (including some tall herbs) on well-drained acid soils.
LW2b 11 250–10 800	Reduced pollen content, increased erosion and possibly stronger winds all indicate climate deterioration (the Younger Dryas).
LW2c 10 800–10 300	Climate amelioration indicated by increased pollen content and spread of dwarf-shrub communities.
LW3 10 300–9 340	Increased tree pollen content (50%) and reduced herb pollen (35%). Increase in tall-herb and fern communities, birch and hazel at expense of dwarf vegetation. Stable zone with almost complete vegetation cover. Evidence of Calluna on higher land with leached soils.
LW4 9 340–3 300	Local taxa dominant. From 9.4 ka BP oak, alder and elm appear (but may be transported pollen). From 8.7 ka BP hazel scrub and bog myrtle increase. From 8.5 ka BP Calluna increases. From 5 ka BP evidence of elm declines. From 4.5 ka BP increase in bog communities suggests wetter conditions. From 4 ka BP evidence of pastures.
LW5 3 300–0	High Calluna content. Reduction in tall herbs. 3 ka BP: evidence of charcoal and pastoral weeds. 2.5 ka BP: convincing evidence of human activity (reduction in trees, increase in grasses and reeds indicates mixed pastoral/arable economy). 1.8 ka BP: appearance of cereals.

Source: Pegler, 1979.

total sunshine, short growing season, frequent storms and salt-laden spray (Peglar, 1979). The only woodland (hazel, birch and willow) found in Caithness today occurs in sheltered valleys and gorges in the southeast. However, it is likely that present-day agricultural and economic activities, such as sheep grazing, are major contributing factors preventing the establishment of extensive woodlands.

The major findings of these two studies are confirmed in a concise review of Scottish Quaternary pollen records (Walker, 1984). Walker discusses the similarities and differences between the Caithness/Orkney region and the rest of Scotland. Over 60 Scottish pollen records of the Late-glacial are available and are reviewed in more detail by Gray and Lowe (1977).

The earliest pioneer vegetation in Scotland developed on bare mineral substrates from about 13 ka BP onwards, and was succeeded by a more complex and spatially variable pattern of vegetation, reflecting variations in environmental factors such as aspect, altitude, exposure and soil type. There is evidence across much of Scotland for climate deterioration in the Loch Lomond Stadial, although the precise chronology is often uncertain.

In the Holocene, regional differences emerge in the extent and type of woodlands. Birch, for example, reached central and southern Scotland before 10 ka BP, but did not reach the north until after 9 ka BP. Pine and juniper failed to expand into Caithness and the Orkney islands. These regions never saw the mixed deciduous forests which were established in southern Scotland from 8.5 ka BP onwards. From about 5 ka BP, anthropogenic effects dominate the pollen record and climate inferences cannot be made. The date at which the first convincing evidence of human activity is seen does, however, vary across Scotland. In Aberdeenshire, for example, evidence of deforestation is seen before 5 ka BP, whereas none is found in the Cairngorms until 3.5 ka BP. In Caithness, convincing evidence has a relatively late date, at about 2.5 ka BP.

Although pine pollen has been found in Caithness cores, there is no direct evidence (such as fossil tree stumps) that pines actually grew in the area (Peglar, 1979). The most likely origin of this pollen is wind-blown transport. It is known that pine did not grow in Scotland until about 6.7 ka BP so that any occurrence of pollen before that date must be due to transport. It has been proposed that the presence of wind-blown pine pollen can be used as an indicator of climate and atmospheric circulation (Tipping, 1989). Tipping speculates that the rapid southward shift of the polar front at the end of the Younger Dryas (about 10.3 ka BP), and the associated weakening of the atmospheric circulation system over the British Isles, led to an increase in the transport of pine by winds from the south. The reasoning is that, as the atmospheric circulation weakens, so westerlies become less dominant, and the meridional (south–north) circulation becomes more important.

All the evidence discussed so far comes from the Holocene. A few pollen records are available for interglacial episodes prior to the Late Devensian glacial period. These come from coastal and island sites: Fugla Ness and Sel Ayre on Shetland; St Kilda, Toa Galson and Tolsta Head on Lewis; and, on the east coast, Teindland, Kirkhill, Tipperty and Benholm. The most detailed records are from the Shetland sites. The evidence indicates relatively low tree-pollen content, but is limited and often ambiguous. No other proxy data sources are available for these earlier interglacials. If, however, these pollen records are assumed to be representative, then it appears that the northern fringes of the British Isles had a more open landscape than southern England, where extensive woodlands developed (Walker, 1984).

Pollen records from the Cumbrian lowlands

The most comprehensive study of late Quaternary vegetational history in the Cumbrian lowlands is that of Walker (1966). Eight records are available from sites at Scaleby Moss, Oulton Moss, Moorthwaite Moss, Abbot Moss, Bowness Common, Glasson Moss, St Bees and Ehenside Tarn. Walker has developed a composite record from the Late Devensian to the present which is subdivided into 23 zones, C1–C23. These zones can be classified into three major periods: Pioneer Vegetation, zones C1–C9 (about 14 ka BP to about 8.5 ka BP); Post-Glacial Forest Period, zones C10–C15 (about 8.5 ka BP to 6 ka BP); and Forest Reduction, zones C16–C23 (about 6 ka BP to date).

Table 7.7 Major vegetation characteristics of 15 zones identified in Cumbrian lowland pollen cores. The zones are numbered from C1, the oldest, to C15, the most recent.

Zone	Vegetation characteristics
Pioneer vegetation about 14–8.5 ka BP	
C1–C4	Predominantly herbaceous, open-tundra/steppe-type vegetation with occasional birch/pine in places with good aspect and soil, more extensive dwarf birch/willow shrubs. More diverse/undifferentiated in C3/C4.
C5–C6	Late-glacial, Alleröd interstadial. Expansion of wooded areas (birch and some pine). More stable communities, even limited areas of closed woodland. Equilibrium developed in 3–4 ka, briefly maintained, then broke down.
C7–C8	Younger Dryas. Pronounced instability with reduction in trees including pine and birch. Increases in herbs favouring damp soils. Towards end some recovery of birch and other trees.
C9	Post-glacial period. Differential rates of expansion of birch woodland across region. Patchy vegetation, open-canopy birch woodlands with juniper, hazel along margins and in clearings. Grasslands and herb-rich communities in less stable soil.
Post-glacial forest period about 8.5–6 ka BP	
C10–C13	Rapid expansion of hazel into birch woodlands, followed by elm and oak. Reappearance of shade intolerant dry-land herbs. Individual sites dominated by local conditions. Elm and oak arrived about same time but elm colonized more rapidly in association with decline in birch and hazel. Later, oak increased more rapidly, hazel declined further, birch reached steady state (8.2–7.8 ka BP). Stable climax community never reached.
C14	Elm severely reduced, very little pine.
C15	Reappearance of shade-intolerant dry land herbs. Forests less continuous.

Source: Walker, 1966.

143

Table 7.7 shows the major vegetation characteristics associated with zones C1–C15. After this time, anthropogenic influences dominate. Walker attempts to interpret the pollen cores in terms of quantitative climatic parameters using present-day distributions and the known tolerances of different species. Here we summarize Walker's assessment of the post-glacial vegetation succession in very general climatic terms.

The vegetation associated with zones C1–C9 indicates low summer temperatures, but not high arctic conditions. Within zones C5 and C6, there is evidence of climate amelioration (between the Late Devensian glaciation and the Younger Dryas, termed the Alleröd interstadial in Europe). More severe conditions in zones C7 and C8 are associated with the Younger Dryas stadial. From this point onwards, the climate improves and becomes more oceanic. Conditions from the Late-glacial (C10–C15) onwards are relatively wet, particularly at about 7.3 ka BP. Walker concludes that most of the changes in the Pioneer Vegetation and Post-Glacial Forest periods can be explained by independent self-generated development under oceanic climate conditions.

The Forest Reduction period (zones C16–C23, about 6 ka BP to date) is dominated by anthropogenic influences. Early forest clearances accentuate the natural elm decline and the increase in birch and hazel. The pollen records reflect changes in agriculture. From about 6 ka BP onwards there is evidence of clearance and grazing on the good coastal soils. Cereals appear over most of the region in the period from about 3.75 ka to 3.4 ka BP. The reduction in trees is accompanied by an increase in herb communities and by blanket bog development.

These changes in lowland Cumbria are compared with the vegetational history of the Lake District mountains and the Morecambe Bay lowlands in a later review (Pennington, 1970). It is concluded that the region of northwest England has lain close to the altitudinal and latitudinal limit of continuous woodland throughout the post-glacial period. Lime, for example, reaches its northernmost limit in the area. Pennington also concludes that no climatic influence on vegetation can be distinguished after 5 ka BP because of strong anthropogenic influences.

Pollen records from Wales
The majority of pollen records from Wales come from the Late-glacial and Holocene period. It has, however, been suggested that the interglacial deposits at West Angle (Dyfed, south Wales) are of Hoxnian age (Stevenson and Moore, 1982). Four pollen assemblages dominated by temperate forest taxa are identified. Their general features show a greater resemblance to Hoxnian sequences from eastern England than to Ipswichian sequences. Fir, for example, is recorded at the West Angle and Hoxnian sites, but not at the Ipswichian sites.

Most Welsh pollen-based stratigraphies of the Late-glacial/early Holocene period show a similar pattern (Hibbert and Switsur, 1976; Lowe, 1981; Walker, 1982; Heyworth *et al.*, 1985). Open-habitat vegetation is succeeded by more stable vegetation. Initially this is characterised by juniper and later, prior to the Younger Dryas, by birch. The vegetation changes are accompanied by a transition from minerogenic to organic sediments until the onset of the Younger Dryas, when minerogenic material increases (Walker, 1982).

Some of the uncertainties associated with the dating of pollen records have been overcome by Walker and Harkness (1990), in an analysis of a sedimentary sequence from Llanilid (Mid Glamorgan, south Wales). Radiocarbon dates were obtained for samples from a section through Late-glacial and early Holocene sediments. In order to identify and correct for potential contamination problems, the material was divided into alkali soluble (humic) and alkali insoluble (humin) components, and the estimated dates compared (Walker and Harkness, 1990). Eleven pollen assemblages were identified in the section, and radiocarbon dates taken at 12 points

Table 7.8 Chronology of Llanilid, south Wales, sedimentary and pollen records.

Beginning of Late-glacial interstadial *sensu stricto*: 13.2 ka BP

   marks onset of organic sedimentation

Juniper maximum: 12.5–12.4 ka BP

   taken to reflect warmest phase of Late glacial period

Juniper decline: 12.2 ka BP

   phase of juniper decline, succeeded by Gramineae (reeds/wild grasses) episode before expansion of birch, so taken as indicator of climatic deterioration

Birch rise: 11.7 ka BP

   marks abrupt establishment of closed birch woodland during a time when a trend towards cooler summers and colder winters was already established

Birch decline: 11.4 ka BP

   some uncertainty in chronology in this part of the time scale; suggestion of possible climatic deterioration prior to 11.4 ka BP, with short-lived recovery about 100 a later

Close of Late glacial interstadial *sensu stricto*: 11 ka BP

Beginning of Flandrian/Holocene interglacial: 10 ka BP

   marked by fall in Artemesia (herb types) and rise in Juniper

Juniper maximum: 9.85 ka BP

Birch rise: 9.6 ka BP

Hazel rise: 9.3 ka BP

Source: Walker and Harkness, 1990.

The resulting chronology is summarised in Table 7.8 and has been compared with other Welsh and English stratigraphies, and with ocean core records (Walker and Harkness, 1990). In comparison with other records, some small discrepancies are found in dates for the end of the Late-glacial interstadial (the Alleröd) and the end of the Younger Dryas. It is possible to distinguish between spatially consistent events, such as the early juniper maximum; spatially inconsistent events, such as the close of the Late-glacial interstadial; and spatially transgressive events, such as the rise in hazel. It is suggested that hazel migrated rapidly northwards into southern and western Britain, between 10 ka and 9 ka BP, reaching south Wales about 9.3 ka BP (Walker and Harkness, 1990) and north Wales shortly afterwards (Hibbert and Switsur, 1976).

7.3 Ocean records

Ocean core records, such as the orbitally tuned SPECMAP series of Imbrie *et al.* (1984), have provided the global stratigraphic framework for Quaternary research. In addition, spectral analysis of deep-sea cores has allowed the identification of global climate responses at the Milankovitch frequencies (see Section 2.2. and, among others, Shackleton and Opdyke, 1973; 1976; Hays *et al.*, 1976; Berger *et al.*, 1984). Here, we are concerned with the use of records from the oceans around the British Isles as indicators of climate conditions on land.

7.3.1 The North Sea

Sedimentary records from the North Sea provide information on the extent of the British and Scandinavian ice sheets during previous glacial episodes, and the accompanying changes in sea level. In the Anglian glaciation the Scandinavian ice sheet extended across the North Sea to East Anglia, acting as a barrier and diverting the Scottish ice sheet towards the area now occupied by the Wash (Bowen *et al.*, 1986). North Sea sedimentary and geomorphological evidence from the Late Devensian glacial indicates that the land-based British ice sheets met the marine ice sheet some tens of kilometres off the east coast of Scotland (Bowen *et al.*, 1986). The southern central North Sea was dry land at this time and the northern central North Sea was occupied by a shallow arctic sea, probably with a thick sea-ice cover. The Scandinavian ice sheet formed the northern boundary of this shallow sea.

The stratigraphy of North Sea Quaternary sediments is still not known in detail, but recent studies have begun to examine the foraminiferal content of cores. Jensen and Knudsen (1988) have analysed the foraminiferal stratigraphy of a 140 m long core from the central North Sea referred to as boring 81/29. Fourteen foraminiferal assemblages are identified in the core, most of which are representative of cold arctic and subarctic periods characterised by lower sea level and higher salinity than the present day.

Changes in patterns of sedimentation and erosion mean that the record is not continuous, but Jensen and Knudsen have used amino acid and thermoluminescence dates, together with correlations with other North Sea cores, to develop a possible core chronology (Figure 7.5).

7.3.2 The North Atlantic

The Quaternary sediment drifts of the Northeast Atlantic have provided a number of continuous deep-sea core records which are, for the reasons outlined in Section 7.1, valuable indicators of past climate conditions in the British Isles.

Ruddiman and McIntyre (1976) used 16 Northeast Atlantic cores to study climate and water-mass changes over the last 600 ka. They reconstructed water-mass movements along a north–south transect in the Northeast Atlantic Ocean from about 57°N to 42°N. The movement of polar, sub-polar, transitional and subtropical water masses can be traced. At the Last Glacial Maximum (the Late Devensian, about 18 ka BP), polar water masses extended south to 42°N. At the height of the previous interglacial (the Ipswichian, about 120 ka BP) subtropical water masses reached 55°N. Seven major cycles of polar water advance and retreat (glacial–interglacial) cycles can be identified in the 600 ka long record. It is estimated that these water masses migrated across more than 20° of latitude, accounting for sea-surface temperature changes of at least 12°C (Ruddiman and McIntyre, 1976).

Ruddiman and McIntyre also infer surface currents from the ocean core records. In the Last Glacial Maximum, for example, a strong cyclonic sub-polar gyre is thought to have existed off the British Isles.

While the earliest deep-sea core studies concentrated on fossil assemblages as indicators of water-mass types, and on oxygen isotope ratios as indicators of global ice volume, more recent studies have looked at different ecological and chemical aspects of the North Atlantic records.

Analysis of coarse-grained ice-rafted debris and planktonic foraminifera identifies a pattern of cyclic ice-rafting in the Northeast Atlantic over the last 130 ka, with an average periodicity of 11 ± 1 ka (Heinrich, 1988). It is speculated that this pattern is related to the 23 ka precession cycle. When insolation has a summer maximum, meltwater production may reduce salinity, giving increased production of winter sea ice. At the opposite phase of the precession cycle, when insolation has a winter maximum, enhanced iceberg production may again reduce salinity and lead to increased sea ice (Heinrich, 1988).

Benthic Cd/Ca and $^{13}C/^{12}C$ ratios are an indicator of nutrient availability (see Section 2.4). Calcium carbonate percentages reflect processes of carbonate production, dissolution and dilution and can, therefore, be used as an indicator of sedimentation, glaciation and continental erosion (Keigwin and Jones, 1989). Changes in deep-sea circulation and in deep-water production can be identified from regional patterns of sedimentation, including variation in clay mineralogy and tephra content (see Section 2.4).

Figure 7.5 Possible core chronology based on the 140 m North Sea Boring 81/29, developed by Jensen and Knudsen (1988). Fourteen foraminiferal assemblages are identified. Present-day water depth is 88.3 m.

MAJOR HIATUSES	TL DATES	ASSEMBLAGE	OCEAN/CLIMATE CONDITIONS	
		1a 1b	Interglacial conditions, present-day salinity, deeper than 1b. Boreal-arctic conditions, present-day salinity, shallow water.	
		2	Arctic marine conditions, ameliorating to subarctic, hyposaline/close to present day, <50 m water depth.	
		3	High boreal content, possibly interstadial/interglacial conditions.	MIDDLE PLEISTOCENE
→	182±18 ka ↓	4 4a 4b	Arctic marine conditions, with slight amelioration. Deeper water than 4b (20–50 m), salinity close to present. Shallow water, hyposaline.	MIDDLE PLEISTOCENE
→	210±20 ka ↓	5 5a 5b 5c	Arctic marine conditions. Shallower water, reduced salinity, suggests near-shore deposition. Non-marine deposits, probably deposited very quickly. Probably deeper and more saline than 6.	MIDDLE PLEISTOCENE
→		6	Arctic marine conditions, reduced salinity, water depth <10 m, conditions not very extreme, deposition may have taken place in near-shore (estuary?) environment.	MIDDLE PLEISTOCENE
		7	Arctic marine conditions, shallow water (<10 m?), salinity possibly slightly reduced.	MIDDLE PLEISTOCENE
→		8	Distinctly ameliorated conditions, very shallow water and reduced salinity, a relatively warm interstadial.	EARLY PLEISTOCENE
		9	Similar to 10.	EARLY PLEISTOCENE
		10	High arctic marine conditions, present-day salinity, water depth 20–50 m.	EARLY PLEISTOCENE
		11	Similar to 12, water may have been shallower or salinity lower.	EARLY PLEISTOCENE
		12	Mainly arctic conditions, relatively shallow water (<30 m), slightly reduced salinity.	EARLY PLEISTOCENE
		13	Non-marine conditions or rapid sedimentation in a marine environment?	EARLY PLEISTOCENE
		14	Mainly arctic conditions, water depth <50 m, salinity probably not much reduced from present day.	EARLY PLEISTOCENE

Transfer functions have been developed to estimate sea-surface temperatures (SSTs) from faunal assemblages (CLIMAP, 1976; 1981) and deep-sea water temperatures from $\delta^{18}O$ records (Labeyrie *et al.*, 1987).

Many of these new studies have been important in improving our knowledge of climate processes, feedbacks and cause-and-effect mechanisms, particularly those in operation over the last glacial–interglacial cycle. Changes in deep-water production, nutrient distribution and the thermohaline circulation are among the processes thought to play a role in glaciation and deglaciation. The major climate-related changes in the North Atlantic during glacial episodes, such as the Last Glacial Maximum and Younger Dryas, and the interglacial episodes, such as the Holocene, are summarized in Table 2.5 and are discussed in detail in Section 2.4.3.

As already discussed, reconstructions of Northeast Atlantic SST probably provide the most direct indicator of conditions over the British Isles. The potential impacts of high-latitude North Atlantic SST changes on the climate system are discussed in Chapters 2–4. The best estimates of conditions to the west of the British Isles during the Last Glacial Maximum indicate that SSTs were 7–8°C lower than at the present day (McIntyre *et al.*, 1976; CLIMAP, 1976; 1981; Duplessy *et al.*, 1981; Ruddiman *et al.*, 1986; Bard *et al.*, 1987; Keigwin and Jones, 1989). The temperature decrease was greater in January than in August and greater to the south (McIntyre *et al.*, 1976). January SSTs to the west of the British Isles fell to below 0°C. The polar front is the boundary between sub-polar and subtropical water, and can be defined by the 10°C isotherm. Today, the polar front runs from the Carolina coast of North America, south of Ireland, to the Brittany coast. In the Last Glacial Maximum, it was displaced southwards towards Gibraltar and the meridional temperature gradient was strengthened (CLIMAP, 1981; Keffer *et al.*, 1988). In the Younger Dryas (about 11–10 ka BP), the polar front readvanced southwards and the high-latitude areas of the North Atlantic returned to near full-glacial temperatures (Ruddiman and McIntyre, 1981a).

The estimated departure of North Atlantic glacial SSTs from those of the present day (i.e. −7°C to −8°C) is considerably greater than both the typical global average departure of −2.3°C and the tropical latitude change of less than −2°C (CLIMAP, 1976; 1981). While conditions in the North Atlantic during the last glacial were much colder than the present day, conditions during the previous interglacial, the Ipswichian, were only slightly warmer than at present. February SSTs in the mid-North Atlantic (35–65°N) are estimated to have been only 1–1.5°C warmer than today (CLIMAP, 1984).

In addition to the snapshot-type reconstructions of sea-surface conditions discussed above, a long continuous record of North Atlantic sea-surface changes over the last 1.1 Ma is available (Ruddiman *et al.*, 1986). This record has been obtained by splicing together data from Cores K708-7 (0.68 Ma BP to date) and DSDP-552A (1.2–0.68 Ma BP).

The most striking feature of this long composite record is the increasing amplitude of SST fluctuations in the more recent part of the record. Progressively colder glacial minima are indicated from 850 ka BP onwards. Between 700 ka and 400 ka BP, the interglacial maxima become gradually warmer, the amplitude of the dominant cycle (about 95 ka) increasing by a factor of 4. Since 400 ka BP glacial–interglacial fluctuations have apparently remained large, with temperature differences between glacial minima and interglacial maxima of around 11°C in summer and at least 8°C in winter.

Other significant but progressively weaker cycles are evident, with average periodicities of 54 ka, 41 ka, 31 ka, 23 ka and 19 ka. Ruddiman *et al.* also demonstrate that, over the last 700 ka, a cold ice-filled sub-polar gyre has developed and disappeared in the area of 40–50°N with dominant rhythms of 100 ka and 41 ka.

The analysis of Core DSDP 552A in terms of ^{18}O content has been extended to 3 Ma BP (Shackleton *et al.*, 1988). This core indicates a major climate break at about 2.5 Ma BP, with no evidence of ice-rafting before this time. The chronology is provisional prior to 1.6 Ma BP, but the record indicates the greater dominance of obliquity variability (periodicity about 41 ka) in the Early Pleistocene. Over the last 1 Ma, the strength of the precession (about 23 ka) and eccentricity (about 100 ka) cycles has increased. The record also indicates a trend towards higher glacial isotope values (i.e. towards more severe conditions) but not towards lower interglacial isotope values (i.e. no trend towards milder interglacial conditions) over the last 3 Ma (Shackleton *et al.*, 1988).

7.4 European pollen records

7.4.1 Time series

Long pollen sequences are only preserved where continuous undisturbed sedimentation occurs at sites such as lake hollows. All the available long European pollen records, including Grande Pile in France (Guiot *et al.*, 1989), Padul in southern Spain (Pons and Reille, 1988) and Tenaghi Phillipon in Macedonia (Wijmstra, 1969), come necessarily from areas which have not directly experienced ice-sheet development. In the absence of appropriate records from the British Isles, pollen records from Grande Pile in the French Vosges (Woillard, 1979; Woillard and Mook, 1982) and from the Netherlands (Van der Hammen *et al.*, 1971) must be used as a guide to conditions.

Originally, scientists analysed the Grande Pile record in terms of the percentage of different pollen groups such as 'mixed oak' and 'herb type' (Figure 7.6). This record, and those from Les Échets and Lac du Bouchet, have since been reanalysed in order to reconstruct quantitative records of annual temperature and precipitation, using climate transfer functions based on present-day statistical relationships between climate and vegetation (Guiot *et al.*, 1989; de Beaulieu *et al.*, 1991). The reconstructed time series extend back beyond the time of the last interglacial to 140 ka BP.

The patterns and magnitude of change at the three sites are broadly comparable, particularly for annual temperature. Common episodes can be identified: the Eemian interglacial maximum, the Melisey stadials (cold, dry periods), the St Germain interstadials (warm, moister periods), the Last

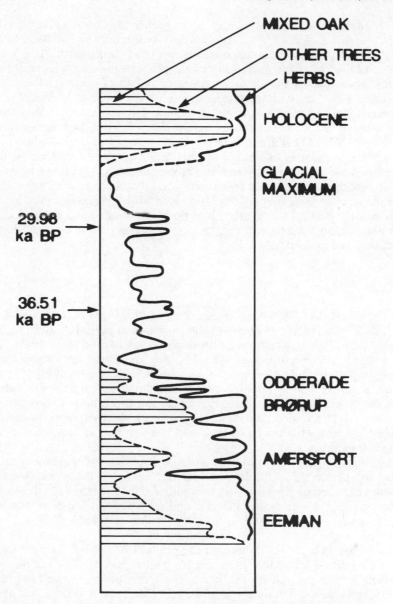

Figure 7.6 Generalized Grande Pile pollen sequence, from 140 ka BP to the present. (After Woillard, 1979.)

Glacial Maximum and the Holocene thermal optimum. For a discussion of the Melisey and St Germain events, see Section 7.5.2. In the Les Échets and Lac du Bouchet records there is some suggestion of a climate deterioration about 11 ka BP, coinciding with the Loch Lomond Stadial in the British Isles (the Younger Dryas). No such event can be seen in the Grande Pile record.

The reconstruction of January and July temperatures for the Lac du Bouchet record reveals changes in seasonality (de Beaulieu *et al.*, 1991). The climate of the Holocene and Eemian interglacials, for example, was more oceanic than that of the St Germain interstadials. However, the errors associated with these estimates are large, in part because of methodological problems, including the lack of perfect modern-day vegetation analogues. The average error associated with the Lac du Bouchet temperature estimates is ±3.4°C (de Beaulieu *et al.*, 1991). In warm periods, however, the maximum error is estimated at only ±1.5°C. The largest errors are, therefore, associated with the temperature estimates for colder periods.

The French sites discussed above provide valuable records of vegetation (and possible climate) changes over time at individual locations. The spatial representativeness of these individual pollen cores is a matter of debate. Spatial patterns of European vegetation change in the post-glacial period are considered in Section 7.4.2.

7.4.2 Spatial patterns

Huntley and Birks (1983) have undertaken a major study of the broad-scale vegetation patterns in Europe over the post-glacial period. Data from 843 representative locations across Europe have been studied and dated by the ^{14}C method. The maps produced by Huntley and Birks tell us something about the pattern and rate of migration of species and the location of glacial refuge areas, as well as possible climatic influences. A number of factors must, however, be taken into account in the interpretation of records from different locations, including: relative pollen productivity and differential pollen transport; effects of site type and characteristics; and reliability of radiocarbon dates.

Maps showing the distribution of individual species and major vegetation units are presented. Huntley and Birks describe these as 'working hypotheses based on the currently available data'. The maps are developed using principal components analysis and knowledge of the present-day ecological and geographical distribution of pollen assemblages.

Isopoll maps showing the distribution of 46 major pollen taxa are produced for up to eleven dates over the period 13–2 ka BP. Differences in the behaviour of individual taxa can be traced. Birch, for example, was probably locally present on the north European plains in the full-glacial period, and reached Britain about 13 ka BP. Hazel reached Britain at about the same time and was present, in some regions, throughout much of the Late-glacial period (i.e. about 15–10 ka BP). This taxon expanded rapidly between 10 ka and 9 ka BP, at a rate of about 1500 m a^{-1}. Alder expanded westwards from the European mountains at rates of about 500–2000 m a^{-1} and had only just reached Britain by 13 ka BP. By 5 ka BP, alder was present across most of the country. In contrast, beech only just reached southeast England by 3 ka BP and fir never arrived at all. This last species expanded relatively slowly into the Alps from 10 ka BP onwards at a rate of

$40-50$ m a^{-1}.

Although the maps of individual taxa give some indication of post-glacial vegetation and climate changes, maps of major ecological units provide a more valuable picture of broad-scale patterns of change. Huntley and Birks define thirteen vegetation units, on the basis of major ecological groupings of taxa (Table 7.9). (It should be noted that some of these units, particularly those found in the Late-glacial period, have no widespread present-day analogue.) They present vegetational maps for ten dates between 13 ka and 2 ka BP. The major characteristics of each map for the period 13–6 ka BP are outlined in Table 7.10. From these maps it is possible to chart progressive changes in vegetation and times of widespread change.

The Late-glacial was characterized by a relatively simple vegetation pattern. At 13 ka BP forests were confined to eastern and southern Spain.

Table 7.9 The 13 vegetation units defined by Huntley and Birks (1983) on the basis of major ecological groupings of taxa. The availability of modern analogues is indicated. Although, for example, modern grasslands exist, these are not generally the natural communities of Huntley and Birks' unit 12.

Vegetation unit	Widespread modern analogue
1) Tundra	yes
1a) xeric variant	no
2) Birch forest	yes
2a) Populus variant	no
2b) xeric variant	no
3) Birch-conifer forest	yes
3a) xeric variant	no
4) Spruce-dominated forest	yes
5) Northern mixed conifer-deciduous forest	yes
5a) pinus variant	no
6) Mixed-deciduous forest	yes
7) Montane mixed conifer-deciduous forest	yes
8) Montane mixed-conifer forest	yes
8a) Pinus/Larix variant	no
9) Mediterranean forest	yes
10) Xeric Mediterranean vegetation	yes
11) Steppe	yes
11a) treeless variant	no
12) Grassland	no
13) Mixed forest	no
13a) xeric variant	no

Table 7.10 Major characteristics of Huntley and Birks' (1983) maps of European vegetational units from 13 ka to 6 ka BP. For key to unit numbers see Table 7.9.

Map	Vegetation characteristics
13 ka BP	Ice sheets extensive in Iceland/Fennoscandia, present in Scotland, Ireland, Pyrenees, Massif Central and Alps. 1 and 11 dominate. 3/4 occur in E. 6 extensive in S Spain.
12 ka BP	Ice-sheet probably absent from Scotland. Ice in Iceland and Fennoscandia may have been more wide spread than at 11 ka BP. Extent of C European ice unclear. 11, especially 11a, dominates S Europe merging into 1 in W. Forests restricted to E wedge (apex C France). 4 extensive in extreme E with 5, 9 and 10 absent. 6 restricted to small area of extreme SE.
11 ka BP	Broad pattern similar to 10.5 ka BP. 2, 3, 4 and 5 increase in extent, as does 6, especially in SW. 11 less extensive in S but 11a predominates. 9 replaced by 10 in S Greece.
10.5 ka BP	1 extends to much of France. In NE 1a, 13a and 2a are widespread. 11 dominates S but 9 is present in Greece. 3 widespread in CN Europe, 13 and 4 occur in CE Europe. 8a is abundant in Alps, 5a occurs locally and 6 very locally. Forests primarily occur within a wedge with apex in N Iberia, extending from Black Sea to NW.
10 ka BP	Ice sheets cover much of Fennoscandia and Iceland. Small ice sheet in W Scotland. Extent of ice in C mountains unclear. 1 extends into N Britain and 3 dominates much of lowland N. 6 occurs locally in W and S, 5a extensive in France. 13 widespread in NE and 4 occurs in CE. 11 and 12 dominate S but 9 locally present in S Greece.
9 ka BP	Ice extensive in Iceland and Fennoscandia. 2 covers much of deglaciated Fennoscandia: 2a, 3 occur locally. 5a widespread in lowland N. 11 absent and 9 only occurs locally in S Greece. 7/8 sparse in Alps and moderately widespread in SE. 6 present in N, more extensive in S.
8 ka BP	Compared with 6 ka BP: 7/8 reduced in C and SE, 11 absent, 5a greater in extent, 4/5 absent from Fennoscandia, and N extent of 6/3 reduced. 2 more extensive than at 6 ka BP in N Fennoscandia and 1 extends S to Iceland.
6 ka BP	Fennoscandia similar to 4 ka BP and present day. 2 may have extended N in E Greenland. In comparison to 4 ka BP and present day: in C and S 7/8 less extensive, no 11 and reduced 9. 7/8 in Pyrenees and S Greece separated from other areas in Alps, etc. 5a only occurs in SE Poland.

To the north of the Alps and other major mountain ranges lay tundra, with steppe vegetation to the south. By 12–11 ka BP the forests had expanded somewhat, although still restricted to a wedge from Russia to the Iberian Peninsula. Otherwise the broad-scale vegetation pattern remained the same. The main Holocene vegetational patterns were established in a period of widespread change between 10 ka and 8 ka BP, which saw the emergence of new vegetational units. Although the major European vegetational units have remained relatively constant since about 8 ka BP, there is unambiguous evidence of widespread disturbance associated with an episode of climate deterioration following the Holocene thermal optimum (i.e. in the period 6–5 ka BP).

Huntley and Birks have attempted to interpret the vegetational unit maps in terms of general climate conditions. They conclude that the Late-glacial was considerably colder than today, particularly in the north, and was probably more arid. There is evidence to suggest that latitude–temperature gradients were steeper than today in the Late-glacial and the mid- to late Holocene (about 6–2 ka BP). Huntley and Birks also consider that the southward movement of the northern limits of mixed-deciduous and northern conifer-deciduous forest in the period 6–5 ka BP reflects deteriorating climate (for example, a shorter growing season, cooler summers, colder winters). In the same period, steppe-type vegetation expanded in southeast Europe, suggesting greater aridity. Mediterranean forest expanded to the north and west.

The rapid changes seen after 10 ka BP may reflect progressive climate changes or, more probably, delayed migration following ice wastage and climate amelioration. It is difficult to interpret the spread of many forest trees solely in climatic terms, but the fast migration of some aquatic plants does support climate amelioration soon after 10 ka BP.

European post-glacial vegetational patterns are strongly dependent on the location of glacial refuge areas and plant migration characteristics. Huntley and Birks identify 17 possible glacial refuge areas at the margins of the continent; in the Late-glacial the richest European taxa are found in the refuge areas located in the southeast. The majority of species migrated from refuge areas at rates of 200–500 m a^{-1}. A maximum rate of 2 km a^{-1} is observed for alder, and a minimum rate of 25 m a^{-1} for manna ash. Elm, sycamore, hazel, hornbeam and pine all migrated at rates in excess of 1 km a^{-1}. The factors which contribute to the location of refuge areas and to differential migration rates are still not fully understood, although it is known that mountains and open sea act as the biggest barriers to migration.

Changes in vegetational units over Britain, taken from the Huntley and Birks maps, are outlined in Table 7.11. Although the broad scale of these maps makes it impossible to discern the regional variability of vegetation identifiable from pollen cores, and outlined in Section 7.2.2, Huntley and Birks' comprehensive study is useful as a general indicator of broad-scale environmental conditions in Europe during the Late-glacial and early to mid-Holocene periods.

The data used by Huntley and Birks (1983) have recently been combined with information from the former USSR (Peterson, 1983) and elsewhere, to produce a multivariate classification of pollen across the entire European continent (Huntley, 1990). The resulting 32 clusters are equated with vegetation units and mapped at 1 ka intervals. Major trends can be identified (Huntley, 1990). At 13 ka BP, only 16 clusters are present, of

Table 7.11 Vegetation units present in Britain from 13 ka BP to present, from Huntley and Birks' (1983) maps.

Map	Vegetation characteristics
13 ka BP	Tundra, with ice in Scotland and Ireland.
12 ka BP	Tundra, no ice present.
11 ka BP	Tundra, with ice cap in Scotland. Birch forest in East Anglia and coastal strip around the Humber estuary.
10.5 ka BP	Tundra, with ice cap in Scotland. Birch-conifer forest in East Anglia.
10 ka BP	Ice cap in Scotland. N England, N Wales, Scotland, Ireland and W Country: tundra. Midlands, S Wales, East Anglia: birch forest. S England: mixed deciduous forest.
9 ka BP	Last remnants of ice sheets in Iceland and Fennoscandinavia. All UK is mixed deciduous forest, except for a narrow strip along the S coast of mixed conifer and deciduous forest.
8–0 ka BP	All UK is mixed deciduous forest.

which five have no modern analogues. The east-west gradient appears to be stronger than at present. The vegetation pattern is much more complex after 11 ka BP, and the east–west gradient is weaker. The extent of no-analogue vegetation units reaches a maximum at 10–9 ka BP. By 8 ka BP, a large number of vegetation units are still present, but the patches of each are generally of greater extent than during previous periods (i.e. a more coherent pattern is evident). The general features of the present-day vegetation distribution are established by about 6 ka BP, although central and southern Europe show greater heterogeneity than is currently observed.

From these maps, Huntley (1990) draws four general conclusions about the nature of vegetation changes. First, vegetation communities are impermanent assemblages of taxa, persisting for up to 1 ka. Second, many past communities have no extensive modern analogues. Third, although the spatial pattern of vegetation types may show some persistence, the taxonomic composition of communities changes in response to continual environmental changes. Deciduous broad-leaved forest taxa have been extensive in northern and western Europe over the last 8 ka, for example, but four different vegetation units have dominated at different times. Fourth, the changes in vegetation patterns and community composition reflect primarily macro-climate change. Huntley argues that, at this broad scale, human influence is not detectable.

7.5 Regional–global and land–ocean correlations

7.5.1 Methodological problems

In order to extend the analysis of past climate conditions in specific regions beyond the period covered by locally available records, it is necessary to explore correlations between the discontinuous, fragmentary and often poorly dated regional land-based data and the continuous, mainly marine-based, global and regional palaeoclimate records. There are, however, a number of factors inherent in the nature of the different palaeoclimate indicators which complicate such intercomparisons. This means that, even within a relatively small region such as the British Isles (Bowen *et al.*, 1986) or the southern North Sea (Gibbard *et al.*, 1991), some difficulty is experienced in correlating events. There are three factors which complicate the comparison. First, the various palaeoclimate indicators have differing spatial and temporal resolutions. Second, the climate system often exhibits lags and spatially transgressive responses to forcing mechanisms as, indeed, do the proxy indicators themselves. Third, there is no reason to expect a consistent regional response to climate forcing. These factors are discussed in detail below.

First, there is the differing spatial and temporal resolution of the various palaeoclimate indicators. To illustrate this, we compare the characteristics of a land-based proxy indicator with the characteristics of the proxy data available from deep-ocean cores.

One of the most sensitive land-based proxy indicators of climate change is fossil Coleoptera. Existing reconstructions in Britain are, however, restricted to a relatively few regions such as lowland England (Coope, 1975; Atkinson *et al.*, 1987). Coleoptera species are very responsive to changing climate and are capable of migrating in response to ameliorating or deteriorating climate at faster rates than plant species. They are not dependent, for example, on soil changes. For these reasons, climate reconstructions based on Coleoptera reveal shorter and possibly more regionally localized fluctuations than those based on fossil pollen.

The greater sensitivity of Coleoptera records is demonstrated by the first study of Late-glacial Coleoptera from southern Europe (Ponel and Coope, 1990). Stratigraphies based on pollen and Coleoptera assemblages from the same site at La Taphanel, in the French Massif Central, are compared. The broad features of both stratigraphies show general agreement, but more detail is seen in the Coleoptera record. A short period of cold climate, ending about 13 ka BP and identified as the Older Dryas, is seen in the Coleoptera record but is almost indistinguishable in the pollen record. The Coleoptera record also indicates much more intense warming, to present temperature levels, at about 13 ka BP and at the start of the Holocene than does the pollen record (Ponel and Coope, 1990).

At the other extreme, deep-sea cores record low-frequency climate changes. The correspondence of fluctuations and events between cores implies that they are representative of global changes. The temporal resolution is much coarser than that of beetle or pollen records. The composite, orbitally tuned SPECMAP $\delta^{18}O$ record has a temporal resolution of 2 ka (Imbrie *et al.*, 1984). Ruddiman and McIntyre's (1976) study of North Atlantic water-mass variations, based on ecological data, has a resolution of 1 ka. Errors in dating must also be taken into account when comparing records. The relatively high-resolution $\delta^{18}O$ record of Martinson *et al.* (1987), for example, has an overall dating error of ±5 ka spread over the 300 ka record.

Problems occur in the identification of the same event, particularly the onset and termination of glaciation, in land-based and ocean records. Marine $\delta^{18}O$ records are conventionally divided into warm (odd-numbered) and cold (even-numbered) stages (Funnell, 1991). The stage boundary dates are arbitrarily set at the mid-point between maximum and minimum isotope values. This means that they cannot be directly correlated with any identifiable datable event on land, such as the first appearance of mixed oak woodland in northwest Europe, which is conventionally taken to mark the start of interglacial conditions (Bowen, 1978).

The second factor which complicates the comparison of palaeoclimate records concerns the response of the climate system to forcing mechanisms and, in turn, the response of the palaeoclimate indicators to the climate change. The system often exhibits lags and spatially transgressive responses. This means that the timing of the response of different proxy indicators to the same event may vary. There may also be differences in the response time of the same proxy indicator with geographic location.

North Atlantic SSTs lag the global ice volume record by varying lengths of time (CLIMAP, 1984). In the transition to the last interglacial at about 125 ka BP, the lag was 1–2 ka, increasing to 6–7 ka in the transition from the last interglacial maximum. Emerging evidence indicates that there are a variety of lags in the climate system's response to orbital changes. It is apparent that some regions may respond 'out of phase' with respect to other regions (see Chapter 2). In particular, North Atlantic SSTs lag the global ice volume response, whereas Southern Hemisphere SSTs appear to lead the response to orbital forcing (Imbrie *et al.*, 1989; see also Section 2.2).

A spatially transgressive response to processes such as glaciation and deglaciation is often observed (Duplessy *et al.*, 1981). We know, for example, that the Laurentide and Fennoscandinavian ice sheets disintegrated at different rates during the post-glacial period. Deglaciation began relatively early in the European Arctic, and the Fennoscandinavian ice sheet disappeared by 8 ka BP. In contrast, a residual Laurentide ice sheet was still present at 6 ka BP.

It has been suggested that deglaciation began soon after 18 ka BP with the seasonal disappearance of sea ice in the Norwegian and Greenland Seas. Various feedback effects would have been initiated and thus, it is argued, continental ice sheets began to melt some 4 ka before the Holocene insolation maximum which, over the high and middle latitudes of the Northern Hemisphere, occurred at about 11 ka BP (Duplessy *et al.*, 1986). The importance of changes in the heat budget over the high-latitude North Atlantic is further illustrated by marine and terrestrial records which indicate that the two major phases of post-glacial warming (13–11 ka BP and 10 ka BP onwards) were more rapid in Europe than in North America (Overpeck *et al.*, 1989).

Turner and Hannon (1988) have examined the spatially transgressive response of southwestern European climate in the Late-glacial period. Sites in northwest Spain show signs of climate amelioration 0.5–1 ka prior to British sites. The lags in the climate response of the British Isles are, primarily, related to the northward movement of the polar front, from the latitude of the Iberian Peninsula to the vicinity of Iceland. Hence warming in the Pays Basque region lags northwest Spain because of the influence of colder water in the Bay of Biscay. The climatic lags are exaggerated by lags in the vegetation response, which are themselves due to lags in plant migration and soil development (Turner and Hannon, 1988).

Lags in the climate and environmental responses mean, therefore, that absolute synchronism is not essential in order to ascribe individual events to the same causal process. On the other hand, temporal correlation does not necessarily imply a causal link.

Finally, when comparing palaeoclimate records, it should be recognized that there is no reason to expect a consistent response to climate forcing. For example, some areas may cool, even though there is a generalized warming. It is now generally agreed that the world has warmed by about 0.5°C over the present century (Jones *et al.*, 1986). This warming has not, however, been evenly distributed either spatially or temporally. Between 1947 and 1986 large parts of the North Atlantic Ocean, North Pacific, China, northwest Europe, the Mediterranean, Greenland and eastern North America experienced cooling of up to 1.5°C (Jones, 1990). In the same period, parts of the Eurasian continental interior and western North America warmed by at least 1°C. Palaeoclimate reconstructions of global climate conditions may well obscure similar variability in the spatial and temporal distribution of climate change.

7.5.2 *Correlations over the last glacial–interglacial cycle*

The most detailed palaeoclimate records and the most convincing correlations between global and regional, and between land-based and ocean records, have been obtained for the last glacial–interglacial cycle (i.e. the last 125 ka). In order to establish these correlations, it is first necessary to determine the global pattern of change over this period. Five continuous reconstructions representing global changes in the cryosphere, oceans and atmosphere over the last glacial–interglacial cycle are, therefore, presented here.

First, Figure 7.7 shows the composite, orbitally tuned, $\delta^{18}O$ isotope SPECMAP record (Imbrie *et al.*, 1984). This record is representative of global changes in ice volume (Mix and Pisias, 1988). The second record (Figure 7.8), also representative of global changes in ice volume, is the high-resolution orbitally tuned record of Martinson *et al.* (1987). It is based on Core RC11-120 from the southwest sub-polar Indian Ocean. The third record represents ecological water-mass changes in the North Atlantic

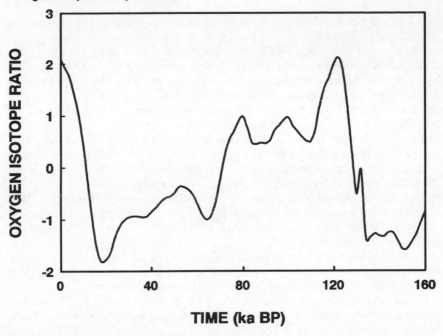

Figure 7.7 SPECMAP composite oxygen isotope record, representative of global ice volume. (Data from Imbrie *et al.*, 1984.)

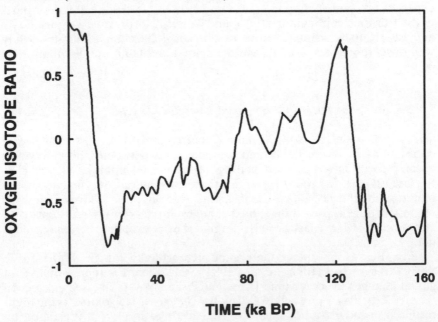

Figure 7.8 Oxygen isotope record from the Indian Ocean Core RC11-120, representative of global ice volume. (Data from Martinson *et al.*, 1987.)

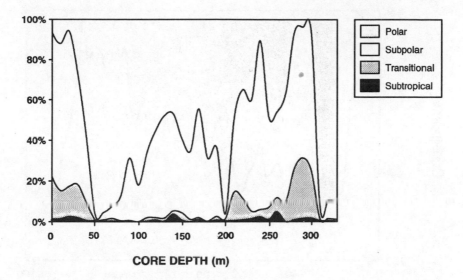

Figure 7.9 Water-mass changes in the North Atlantic over the last glacial–interglacial cycle reconstructed from Core K708-7. (Data from Ruddiman and McIntyre, 1976.)

indicated by analysis of Core K708-7 and is shown in Figure 7.9 (Ruddiman and McIntyre, 1976; Section 7.3.2). The fourth (Figure 7.10) shows changes in the atmospheric concentration of CO_2 estimated from the ancient air bubbles trapped in the Vostok ice core in Antarctica (Barnola *et al.*, 1987). (The potential role of CO_2 in amplifying orbital forcing is discussed in Section 2.4.) Finally, the fifth record (Figure 7.11) is the global sea-level record of Chappell and Shackleton (1986) which is based on analysis of marine $\delta^{18}O$ records and dated coral reef terraces from Barbados.

These five reveal a broadly consistent picture of climate and climate-related change over the last interglacial–glacial cycle. The cycle shows the classic sawtooth pattern: the cooling phase of the cycle is slower and longer than the warming phase. The glacial phases of the cycle (stages 4 and 2 in the $\delta^{18}O$ records) are characterized by high global ice volume, extensive polar water masses in the North Atlantic, low atmospheric CO_2 concentration and low global sea level. Conditions are reversed in the interglacial phases of the cycle (stages 5e and 1 in the marine records).

Closer examination of the various records reveals differences in the detail of each reconstruction and the existence of leads and lags. For example, atmospheric CO_2 falls to similar low concentrations at about 60 ka and 18 ka BP, whereas global ice volume is considerably greater at 18 ka BP than at 60 ka BP. Despite these differences, the general pattern of change is similar in each record.

In Table 7.12 we give the dates of a number of climatic episodes observed on land, mainly over Britain, throughout the last interglacial–glacial cycle. Specific events identified in marine, largely North Atlantic records, are

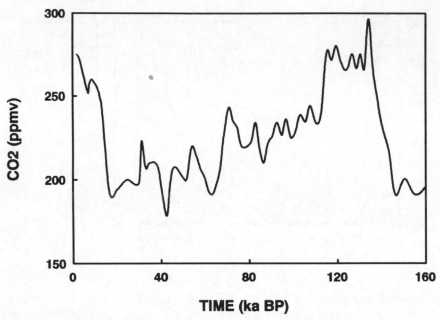

Figure 7.10 Changes in atmospheric CO_2 concentration reconstructed from the Vostok ice core. (Data from Barnola *et al.*, 1987.)

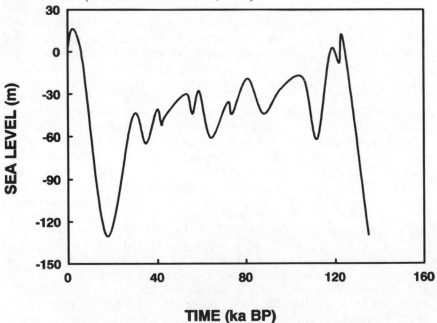

Figure 7.11 Global sea level reconstructed from New Guinea coral reef records. Sea level is shown as departure (m) from present day. (Data from Chappell and Shackleton, 1986.)

Table 7.12 Selected events from the land-based record (mainly Britain/ northwestern Europe).

Reference	Dates	Event
Pennington, 1970 NW England vegetation	26–15 ka BP	Full glaciation
	12380 ± 230 BC	Establishment of pioneer vegetation
	12380–9920 BC	Pre-Alleröd period
	9920 ± 120 BC	Start of Alleröd interstadial
	8878 ± 185 BC	Post-Alleröd, severe periglacial erosion. 50–100 years of cold summers/heavy winter snow
	up to 3 ka BC	Warming, with variable precipitation
	3 ka BC on	Late post-glacial period, elm decline
Lamb, 1977a; 1977b UK/Europe	70–60 ka BP	Establishment of ice sheets
	59–57 ka BP	Chelford interstadial
	50 ka BP	Early maximum of Irish Sea/UK ice
	45–40 ka BP	Upton Warren interstadial
	10.8–10.2 ka BP	Small glaciers in Lake District
	9–8 ka BP	Boreal period
	8–5 ka BP	Atlantic period
	5–3 ka BP	Sub-boreal period
	3–2.5 ka BP	Sub-atlantic period
	AD 1150–1300, late 900s	Little optimum in Europe
Vasari, 1977 NE Scotland	11 850 a BP	Boundary of Bölling stadial/Alleröd interstadial
	10 650 a BP	Boundary of Alleröd/Younger Dryas
	10 145 a BP	Boundary of Younger Dryas/Holocene
Walker, 1980 S Wales	~10.59–9.97 ka BP	Younger Dryas
Lowe, 1981 N Wales	~11.16 ka BP	Onset of Younger Dryas
	~10.04 ka BP	Transition to Holocene
Williams and Wigley, 1983 Europe	AD ~300–700	Period of cooling
	AD ~700–800	Period of cold conditions
	AD 1100–1300	Medieval Optimum
Heyworth et al., 1985 mid-Wales coast	~10.9–10.1 ka BP	Younger Dryas
	~10.55 ka BP	severest conditions
Sutcliffe et al., 1985 N England	238–160, 118–72 and 60–29 ka BP	Dates of interglacial/interstadial episodes estimated from speleothems
	~83 ka BP	Wolverine present in N England

Table 7.12 Selected events from the land-based record (mainly Britain/ northwestern Europe). (Continued)

Reference	Dates	Event
Bowen *et al.*, 1986		ENGLAND:
	26 ka BP onwards	Maximum glacial expansion
	18–17 ka BP	Glacial maximum
	by 14.5 ka BP	Start of retreat
	11 ka BP	LOCH LOMOND ADVANCE
	10.65 ka BP on	Glacial retreat
		SCOTLAND:
	13.5–13.0 ka BP	Start of retreat
	post 10.9 ka BP	Loch Lomond Advance maximum
Stringer *et al.*, 1986		ENGLAND:
	125 ka BP	Ipswichian peak
	83 ka BP	Glacial episode
	81 ka BP	Interstadial
Grove, 1988	AD 1200–1399	Onset of Little Ice Age
	AD 1550–1850	Maximum extent of Little Ice Age
	1600–1610, 1690–1700,	Main European glacial advances
	1770s, *c.*1820, *c.*1850,	
	1880s, 1920s, 1960s	
Jardine *et al.*, 1988	30–29 ka BP	WC Scotland ice-free
	33.5–26 ka BP	Low lying Scotland ice-free
West, 1988	21–13 ka BP	Deposition of glacial till in NW Germany, Denmark, S Sweden, S Finland and E Britain
Cwynar and Watts, 1989	12.6 ka BP	Re-establishment of post-glacial vegetation
Ireland	10.6 ka BP	Onset of Younger Dryas
Dansgaard *et al.*, 1989 Greenland	10 720 ± 150 a BP	Onset of Younger Dryas, which later terminated within 20 a
Gordon *et al.*, 1989 C England	124, 104, 90.5 ka BP	Ipswichian interglacials
	76, 57, 50, 45, 36 and 29 ka BP	Devensian interstadials
Hillaire-Marcel and	~80 ka BP	Early Wisconsin glacial advance
Causse, 1989	~46 ka BP	Significant retreat of Laurentide ice
E Canada	~22 ka BP	Maximum extent of Laurentide ice
Hovan *et al.*, 1989 China/Africa/global?	473, 343, 271, 185, 137, 69 and 28 ka BP	Maximum aeolian flux indicating cold and dry (glacial-type) climate
Peacock *et al.*, 1989 Scotland	10.5–10 ka BP	Maximum extent of Loch Lomond Advance (Younger Dryas)
Rossignol-Strick and	15 ka BP	Termination 1
Planchais, 1989	14 ka BP	Oak expansion
S Europe	post 12 970 ± 180 a BP	Glacial expansion (reduced arboreal vegetation)

Table 7.12 Selected events from the land-based record (mainly Britain/ northwestern Europe). (Continued)

Reference	Dates	Event
Tipping, 1989 SE Scotland	11–10.3 ka BP 10.3 ka BP	Loch Lomond Advance in highlands Southward shift of polar front
Peteet *et al.*, 1990 NE USA	12 290 ±40 a BP 11–10 ka BP	Spruce maximum in NE USA Younger Dryas
Ponel and Coope, 1990, Massif Central, France	~13 ka BP	Onset of rapid warming
Seret *et al.*, 1990 French Vosges	50–30 ka BP	Maximum Devensian ice sheet advance
Walker and Harkness, 1990, S Wales	~13.2 ka BP ~12.5–12.4 ka BP	earliest organic sediments Late-glacial interstadial maximum
Rendell *et al.*, 1991 British Isles	100–90 ka BP	Chelford interstadial

listed in Table 7.13. These tables are not intended to be comprehensive, but illustrate the range of data which is typically available. The extent to which individual events in the land record (Table 7.12) and in the marine record (Table 7.13) can be correlated with continuous marine-based chronologies, such as those shown in Figures 7.7 and 7.8, must be assessed.

The major features of land-based and marine records can be reliably correlated (Bowen *et al.*, 1986; Sibrava, 1986). The present interglacial period (the Holocene) corresponds to stage 1 of the marine oxygen isotope chronology. The last glacial (known as the Late Devensian in the British Isles and as the Late Weichelian elsewhere in Europe) corresponds to stage 2. It is also now generally accepted that the previous interglacial (known as the Ipswichian in the British Isles and as the Eemian elsewhere in Europe) corresponds to stage 5e (Gascoyne *et al.*, 1981; Stringer *et al.*, 1986).

Early studies ascribed the Ipswichian to the whole of stage 5, but a more complex view of this stage is now preferred. The number of associated interglacial, interstadial and stadial peaks is, however, still somewhat controversial. Analysis of long French pollen records, such as Grande Pile, indicates two periods of relatively warm climate (the St Germain I and II interstadials) following the Eemian interglacial (Turon, 1984; Guiot *et al.*, 1989). It is concluded that these interstadials correspond with stages 5c and 5a of the marine record, respectively. The intervening stadials (Melisey I and II) are correlated with stages 5d and 5b, respectively (Guiot *et al.*, 1989). This sequence of events is not, however, observed in all records.

Subdivisions of marine stage 5 (the warm stages 5e, 5c and 5a) can also be detected in the record of global sea-level change. This has been demonstrated by analysis of high sea-level stands which are indicative of

Table 7.13 Estimated dates of selected events from the ocean-based record of the last glacial/interglacial cycle. Most records come from the North Atlantic region.

Reference	Dates (ka BP)	Event
CLIMAP, 1984	127–116	Stage 5e, but response of N Atlantic SST lags
Duplessy *et al.*, 1986 Core CH73-139C	just after 18	NORTH ATLANTIC: Seasonal disappearance of Norwegian/Greenland sea ice
	15.8–13.3	Glacial retreat (Termination 1A)
	13.3–11.0	Alleröd (interstadial)
	12.0–10.5	Pause in warming
	10.5 onwards	Warming (Termination 1B)
Boyle and Keigwin, 1987	11.5–9.0	Increase in Cd/Ca ratio and reduced $^{13}C/^{12}C$ ratio indicate reduction in NADW production
Labeyrie *et al.*, 1987	115	Atlantic deep water cools by 2.5°C
	70	Atlantic deep water cools by 1°C
Broecker *et al.*, 1988 V23-81/Grande Pile	33–25	Interstadial (also seen in USA)
	20–14	Stadial
Jones and Keigwin, 1988	~15	Rapid (*c*.500 a) destruction of Barents Shelf ice sheet, causing up to 15 m eustatic sea-level rise
Bard *et al.*, 1989 $\delta^{18}O/^{14}C$ ages Globeringa bull.	14.5–13.5	Sea level rose 40 m
	12.2	Sea level ~75 m < present
Fairbanks, 1989 Barbados coral reefs	Last Glacial Maximum	Sea level 121 ± 5 m < present
	17.1–12.5	Sea level rose 20 m
	~12	Sea level rose 24 m in < 1 ka (MWPIA)
	~11.0–10.5	Minimum rise
	10.5–10.0	Faster increase
	~9.5	Sea level rose 28 m (MWPIB)
Keigwin and Jones, 1989	13.5, 12.0 and 9	NORTHWEST ATLANTIC: Warm peaks and (for first 2 dates) meltwater discharge peaks
	10.7	Younger Dryas maximum
Overpeck *et al.*, 1989		NORTH ATLANTIC:
	13.0–12.6	Rapid warming
	~11	Younger Dryas
	10 onwards	Warming LAURENTIDE ICE SHEET:
	15–14	Start of retreat
	14–13	Retreat slowed/stopped
	13–11	Peak of retreat

Table 7.13 Estimated dates of selected events from the ocean-based record of the last glacial/interglacial cycle. Most records come from the North Atlantic region. (Continued)

Reference	Dates (ka BP)	Event
Stoker *et al.*, 1989	13–11	Late-glacial interstadial
NE N Atlantic cores	11–10	Younger Dryas
	9.1–9	Onset of Holocene
Bard *et al.*, 1990	13.5 and 11.0	Meltwater pulses
Jansen and Veum, 1990	18 and 13	Lower δ^{13}C suggests enriched
Core V23-81 δ^{18}O		Antarctic Bottom Water, reduced NADW production: triggered by meltwater
	15.2–11.5	Major deglaciation
	10 onwards	2nd deglaciation phase
Ku *et al.*, 1990	120/117, 105, 81	High sea-level stands at Barbados
Lehman *et al.*, 1991	15	Initiation of Fennoscandinavian ice-sheet retreat

interglacial and interstadial conditions. Analysis of coral reef terraces in Barbados indicates three episodes of relatively high sea level (Barbados III, II and I). The uranium-series dates for these terraces have recently been carefully recalculated (Ku *et al.*, 1990). The revised dates are: for Barbados III (two sites), 120 ± 2 ka BP and 117 ± 3 ka BP; for Barbados II, 105 ± 1 ka BP; for Barbados I, 81 ± 2 ka BP. These three high sea-level stands, therefore, correspond to marine stages 5e, 5c and 5a, respectively.

Evidence of high sea-level stands in the Northeast Atlantic coastal regions (Britain, northern France, Belgium, northwest Germany, Denmark, Sweden and Norway) has recently been used to investigate land and marine correlations over the Quaternary period (Bowen and Sykes, 1988; Bowen *et al.*, 1989). A method has been developed by which measurements of time-dependent chemical changes (known as 'isoleucine epimerization') within non-marine mollusc shells deposited in glacial and interglacial sediments can be used to identify periods of high sea level. High sea-level stands are dated using geochronometric methods and the model constrained by the oxygen isotope record of sea level.

Four high sea-level events are identified over the last interglacial–glacial cycle (Bowen and Sykes, 1988). These are correlated with marine stages 5e, 3, the end of 2, and 1. These events are then used to determine the timing and extent of glaciations in the British Isles and Scandinavia (Table 7.14). Three periods of glacial advance are recognized over the last 125 ka (Bowen and Sykes, 1988). Glaciation E is ascribed to stage 4 and is given dates of 70–60 ka BP (but could possibly be associated with stage 6), and is linked to evidence of glacial activity in Scotland (including Orkney and Caithness) and Ireland. Glaciation F in stage 2, about 17 ka BP, corresponds to the

Table 7.14 Proposed chronology of glaciations and interglaciations in the British Isles based on land/sea correlations (Bowen and Sykes, 1988; Bowen *et al.*, 1989). The construction of this chronology is described in the text. The ocean core stages are numbered from 2, the most recent, to 14, the oldest, whereas the high sea-level and glacial events are numbered in the reverse order (from 1, the oldest and from A, the oldest, respectively).

Age (Imbrie *et al.*, 1984) (ka BP)	Ocean core stage	High sea-level event indicating interglacial conditions	Geochronology (ka BP)	Glaciation
		8	11	G Loch Lomond
	2			
		7	17	F late Devensian late Weichselian
24				
	3	6		
59				
	4			E in Scotland and Ireland
71				
	5			
122				
	5e	5	122	
128				
	6			D 'Warthe'
186				
	7	4	191	
245				
	8			C 'Drenthe'? Paviland
303				
	9	3	300	
339			348	
	10			
362				
	11	2	371	
423				
	12			B 'Elster' (Anglian 'Irish Sea')
478				
	13	1		
524				
	14			A 'Elster' ('Elster I'?)
565				

Late Devensian. Glaciation G, also in stage 2, about 11 ka BP, corresponds to the Younger Dryas and Loch Lomond Stadial.

This approach has also been used to develop a detailed correlation between land- and marine-based records of interglacial and interstadial events (Bowen *et al.*, 1989). In this later study, the Upton Warren interstadial has been correlated with marine stage 5a at about 80 ka BP. In an early study of lowland England summer temperatures based on Coleoptera, this interstadial was given dates of about 43–40 ka BP (Coope, 1975). Coope also identified a Chelford interstadial centred on 60 ka BP and a Windermere interstadial at about 12 ka BP. More recently, thermoluminescence dating of the Chelford Sands formation has given a date of 100–90 ka BP to this event (Rendell *et al.*, 1991). The Windermere interstadial corresponds to the well-documented post-glacial climate amelioration which occurred before the Younger Dryas reversal (Table 7.12). Analysis of speleothem growth frequency in Britain indicates at least six interstadial and three interglacial peaks over the last 125 ka (Gordon *et al.*, 1989; Section 7.2.2). The speleothem records do not, however, reveal anything about the relative magnitude of these events and it is not possible to correlate them with previously identified interstadial episodes such as the Chelford or Upton Warren.

The above discussion illustrates some of the problems experienced in trying to correlate land and marine records. We can be certain that a number of relatively warm (interstadial) and relatively cool (stadial) episodes have occurred in the British Isles and northwestern Europe over the last interglacial–glacial cycle. We cannot, however, be certain about the precise chronology of these events.

Some of the problems encountered in developing land–marine correlations for the British Isles arise because the most recent ice advances have effectively wiped out evidence of previous climate episodes. We might, therefore, expect it to be easier to develop correlations in regions beyond the ice-sheet limits.

In southern Europe $\delta^{18}O$ records from the Tyrrhenean Sea have been correlated with vegetation changes in Italy (Rossignol-Strick and Planchais, 1989). Between 55 ka and 9 ka BP, a strong negative correlation exists between oxygen isotope ratios and the abundance of deciduous oak pollen in southern Italy and Sicily. The occurrence of oak is mainly constrained by moisture availability, hence the pollen record indicates drier conditions in glacial episodes, which may or may not have been associated with colder temperatures. The fluctuations of grass, oak, Artemesia and fir pollen abundances appear to be phase-locked with the $\delta^{18}O$ record (Rossignol-Strick and Planchais, 1989). Lags can, however, be identified. Glacial Termination I in the marine record, for example, is dated at 15 ka BP while oak expands in Italy and Greece after the slightly later date of 14 ka BP.

General agreement has been found between a 350 ka climate record reconstructed from the loess sequence at Achenheim (Alsace, France) and the SPECMAP marine record (Rousseau and Puissegur, 1990). The climate record used in the comparison is based on the first factor calculated in a multivariate analysis of the mollusc assemblages found in the loess. It is considered to reflect temperature primarily. Common features, such as the Younger Dryas event, can be identified in the mollusc and SPECMAP records. The entire loess section contains evidence of at least five glacial cycles, but only the three most recent can be reconstructed in detail (Rousseau and Puissegur, 1990).

In latitudes such as northwestern Europe, glacial–interglacial cycles are characterized primarily by changes in the temperature regime and by the expansion and decay of ice sheets. In other parts of the world these major climatic cycles are expressed principally by changes in the hydrological regime and in the atmospheric circulation. Fluvial, rather than glacial, cycles have been identified in parts of Africa and the western USA (Smith, 1984; Kutzbach and Street-Perrott, 1985; Gasse *et al.*, 1989). There is convincing evidence that the Indian monsoon was weaker during the Last Glacial Maximum and stronger than today during the Holocene thermal optimum (Kutzbach and Guetter, 1984; COHMAP, 1988).

In two recent studies, correlations have been established between oceanic and continental pollen records of the last glacial–interglacial cycle from North Africa (Lezine and Hooghiemstra, 1990; Lezine and Casanova, 1991). In the first (Lezine and Hooghiemstra, 1990), a strong correlation is found between the continental pollen record from Senegal and Mauritania and the pollen content of an offshore marine core during the last glacial–interglacial transition. It is concluded that, in this region of northwest Africa, pollen records from offshore marine cores may be used as an indicator of latitudinal shifts in the vegetation zones on the adjacent continent. The sensitivity of the marine cores is, however, less than that of the continental records. In the second study, the pollen record from the East Atlantic Ocean Core V22-196 is correlated with continental pollen records and with records of high lake levels from North Africa (Lezine and Casanova, 1991). In interglacial periods, the pollen input to the marine core from the humid vegetation zone increases and lake levels rise. Five humid phases are identified at 140–118 ka, 105–96 ka, 92–73 ka, 52–44 ka and 12–2 ka BP. These phases correlate with oxygen isotope stages 5e, 5c, 5a, 3 and 1, respectively.

All the available evidence indicates that the previous interglacial (the Ipswichian), the last glaciation (the Late Devensian) and the present interglacial (the Holocene or Flandrian) have been accompanied by world-wide climate and climate-related changes. However, the spatial extent of other events such as the Younger Dryas and Little Ice Age, observed over the last 125 ka, is debatable. Even if the Younger Dryas and Little Ice Age are assumed to be global events, their short length means that we would not necessarily expect evidence of them in the marine oxygen isotope records, which generally have a relatively coarse temporal resolution. The problems of scale and the causes of short-term climate variability are discussed in more detail in Chapter 3.

7.5.3 Correlations over the Quaternary period

We have established that the main features of the last glacial–interglacial cycle, as recorded on land and in the oceans, present a broadly consistent and coherent picture of climate change. Do these relationships hold for previous cycles?

This question can only be approached by examining long continental records of climate (Kukla, 1989). Such records are rare, and we now consider evidence from around the world before focusing on the implications for northwest Europe, and Britain in particular. The most useful records for this purpose are long pollen records, notably those from Macedonia and the Netherlands (Wijmstra, 1969; Jong, 1988) and Chinese loess stratigraphic records (Hovan et al., 1989; Kukla, 1989; Kukla and An, 1989).

The 900 ka Macedonian pollen record from Tenaghi Phillipon confirms the major stratigraphic features of the marine and Chinese loess records (Wijmstra, 1969; Kukla, 1989). The record indicates periods of cool and dry conditions correlating with isotope stages 22, 16, 12, 6, 4–2. Predominantly milder conditions are associated with stages 15–13 and 11–7.

Jong (1988) has reviewed the evidence for vegetational changes in the Netherlands and northwest Europe over the last 3 Ma. A tentative correlation with the oxygen isotope record of Shackleton and Opdyke (1976) is proposed. A number of uncertainties are, however, encountered. It is not clear, for example, whether the Hoxnian interglacial period should be ascribed to isotope stage 9 or 7. Considerable differences are observed between the vegetation cycles of early interglacial periods such as the Tiglian and Waalian and those in the Late Pleistocene. A number of tree species, for example, became locally extinct after the Early/Middle Pleistocene. These differences may be due to the fact that temperatures were higher during the earlier Quaternary glaciations than at the Last Glacial Maximum (Jong, 1988).

Evidence from a long continuous pollen core from Israel has recently been correlated with the marine oxygen isotope record (Horowitz, 1989). In this region, glacial episodes are characterized by lower temperatures and higher rainfall. The interglacial periods are typically hot and dry. The 3.5 Ma Israeli pollen record indicates that the first major cooling occurred at about 2.6–2.4 Ma BP (Horowitz, 1989). This is in agreement with the Dutch vegetation record (Jong, 1988), the Chinese loess records (Kukla, 1989) and the North Atlantic SST record (Ruddiman et al., 1986), all of which indicate the onset of glacial conditions at about 2.5 Ma BP. This date is conventionally taken as the beginning of the Quaternary period.

The most convincing evidence for land–marine correlations over previous interglacial–glacial cycles comes from Chinese loess records. Loess is a terrestrial wind-blown silt deposit (see discussion in Section 2.2). It has been demonstrated that the main source of the Chinese deposits lies upwind over central China (Hovan et al, 1989). Loess layers in the Chinese sedimentary sequences can, therefore, be used as an indicator of cold and dry climate conditions over the continental interior. The sedimentary layers are dated using magnetic susceptibility records.

The Chinese loess record extends back over the last 2.5 Ma (Kukla, 1989; Kukla and An, 1989). At least 44 glacial–interglacial type shifts can be identified. Ten or twelve of these cycles are particularly strong: the eight strongest are equivalent, both in length and strength, to the most recent cycle. Eleven loess layers indicate prolonged episodes (greater than 40 ka) of glacial-type conditions. The frequency of cycles is greater in the early part of the record (2.4–0.5 Ma BP) in comparison with the last 0.5 Ma. The 'classic' approximate 100 ka sawtooth cycle is only seen during this most recent period.

Correlations have been established between loess layers and even-numbered (cold) isotope stages (Hovan *et al.*, 1989; Kukla and An, 1989). The loess layers dated 650–610 ka BP, 480–430 ka BP, 180–130 ka BP and 75–11 ka BP correlate with isotope stages 16, 12, 6 and 4–2 respectively and, therefore, to the four major European glaciations (Kukla and An, 1989).

The evidence for Quaternary land–marine climate correlations was reviewed at a recent International Quaternary Association conference. In the Introduction to the *Conference Proceedings* Kukla (1989) concludes that we now have sufficient palaeoclimate evidence to confirm that, except for two events at about 1.6 Ma and 1.4 Ma BP, the glacial-type episodes identified in Chinese loess records and in deep-sea cores were accompanied by the major build-up of land-based ice. The reasons for the apparent discrepancies at 1.6 and 1.4 Ma BP are not known.

In a review of Quaternary glaciation in the USA (Fullerton and Richmond, 1986; Richmond and Fullerton, 1986), a provisional chronology is presented which demonstrates a one-to-one match of 11 episodes of continental glaciation with eleven 'cold' stages and substages in the $\delta^{18}O$ marine record for the last 900 ka. Although the number of events matches, the two records are not in phase. It is estimated that the marine record may lag the continental record of glaciation by 0.5–3 ka (Mix and Ruddiman, 1984). The authors emphasize that the maximum of the last glaciation in the USA (the late Wisconsin) did not occur at 18 ka BP, which is conventionally taken as the date of the Last Glacial Maximum. The evidence suggests that it occurred some 2–3 ka earlier (Fullerton and Richmond, 1986).

As in Europe, correlations have been established between US regional records of Late-glacial/Holocene climate change and larger-scale/global events (Delcourt and Delcourt, 1984; Heusser *et al.*, 1985; Peteet *et al.*, 1990). One of the longest pollen records from the USA comes from the Humptulips mire in Washington State (Heusser and Heusser, 1990). This record is about 350 ka long and covers the last two interglacial periods and two periods of glaciation (the Wisconsin and Illinoian). Fourteen pollen zones are identified, the oldest of which is provisionally correlated with oxygen isotope stage 9. The pollen record shows general agreement, through to oxygen isotope stage 5e, with the record from marine core Y772-111 in the Northeast Pacific Ocean, just off the Oregon coast (Heusser and Heusser, 1990).

Is it possible to link land-based events in Europe with the marine Quaternary record? Bowen and Sykes (1988) have used indirect evidence of sea-level change to equate northwest European events with the marine record back to isotope stage 15. (Their methodology was explained in Section 7.5.2.) The marine isotope record indicates that there should have been seven glacial (even-numbered) stages and eight interglacial/interstadial (odd-numbered) stages within this period. Their proposed chronology does indeed contain seven glacial episodes and eight high sea-level events (indicative of interglacial and interstadial conditions), as summarized in Table 7.14. However, there is no evidence of a European glaciation associated with isotope stage 10, and two glaciations are associated with stage 2.

In Britain, the picture is even more disjointed. In addition to the missing stage 10 glaciation, no reliable evidence exists for glaciation associated with stage 14 (Glaciation A), and the evidence for glaciation in stage 6 (Glaciation D) is ambiguous. In the case of Scotland, direct and reliable evidence only exists for four glacial episodes.

Despite the fragmentary nature of the evidence, this analysis indicates that the majority of glacial–interglacial cycles seen in the marine record over the last 500 ka or so have been accompanied by the expansion and disintegration of ice sheets over Britain and the European continent.

7.6 Conclusions

In this chapter, we have identified the general pattern and range of climate change experienced in the British Isles over the Quaternary period. The evidence discussed here demonstrates the extent to which it is possible to relate regional patterns of climate change to larger-scale and long-term forcing mechanisms. We have established that the main features of the last glacial–interglacial cycle, as recorded on land and in the oceans, present a broadly consistent and coherent picture of climate change. Common events, such as the Last Glacial Maximum (18 ka BP) and the Younger Dryas (11 ka BP) can be identified in palaeoclimate records from our three study areas and in longer, continuous records from the oceans and from the European continent. Over the last 500 ka or so, the majority of glacial–interglacial cycles identified in the marine record can be linked with cycles of ice-sheet development and decay in Britain.

8 Climate reconstructions of the last glacial–interglacial cycle

8.1 Introduction

The evidence discussed in the previous chapter provides the general background pattern of climate change over the Quaternary in the British Isles, including the selected study areas (Caithness, Cumbria and Wales). The aim of this chapter is to provide a more detailed picture of the climatic conditions experienced in these regions over the last glacial–interglacial cycle. This period is chosen simply because it is the most recent and, therefore, has been studied in greatest detail. Particular consideration is given to the extreme glacial and interglacial stages. Two different sources of evidence are used to investigate both the magnitude of climate change and the prevailing general circulation patterns. The first is general circulation model (GCM) output and the second is palaeoclimate data.

8.2 Reconstructions based on general circulation models

The features of GCMs and their shortcomings are discussed at some length in Chapter 4. Their use in palaeoclimate studies has recently been reviewed by Street-Perrott (1991). GCM palaeoclimate simulations fall into two categories: snapshot and sensitivity studies.

Climate conditions and circulation patterns at selected times in the past have been investigated through snapshot experiments. For model validation, the output from snapshot simulations must be compared with climate reconstructions based on palaeoclimate data. Model boundary conditions, including the location and size of ice sheets and sea-surface temperature (SST), are prescribed from the palaeoclimate data. Hence model experiments are mainly restricted to comparatively recent and distinctive episodes, such as the Last Glacial Maximum (18 ka BP) and the Holocene thermal optimum (6 ka BP), for which sufficient data are available. Snapshot experiments are able to reproduce many of the general features of the palaeoclimate data and can provide useful information on temperature, precipitation and atmospheric circulation changes associated with particular orbital configurations and boundary conditions (Street-Perrott, 1991).

The many shortcomings of GCMs described in Chapter 4 also apply to the palaeoclimate experiments described here. As in the case of the $2\times CO_2$ equilibrium model studies, the relative level of confidence in model output decreases from the global to the regional scale. This point is illustrated in Table 8.1 for the Last Glacial Maximum. In fact, relative confidence in results for the Last Glacial Maximum on scales down to the broad regional

174

Table 8.1 GCM results from snapshot simulations of the Last Glacial Maximum (18 ka BP) and an estimate of the relative confidence that can be placed in these results.

Model results	Confidence
Global scale	
Surface cooling	High
Reduced intensity of the hydrological cycle	High
Zonal–regional scale (Northern Hemisphere)	
Increased N–S temperature gradient	High
Higher surface pressure over ice sheets (development of glacial anticyclones)	High
Weaker monsoon	High
Intensification of westerly jet stream	Moderate
Splitting of jet stream round Laurentide ice sheet	Moderate
Southward shift of surface storm tracks across the North Atlantic	Moderate
European drying	Moderate
Increased European seasonality	Moderate
Greater/less interannual variability	Unknown
Changes in cloud cover	Unknown

level (for example, the northeast USA) is greater than for $2\times CO_2$ simulations (see Table 4.4). This is because model output can be validated by the palaeoclimate data, such as pollen records (Kutzbach and Wright, 1985). We have no such basis for the validation of $2\times CO_2$ simulations. Furthermore, in most GCM palaeoclimatic experiments, the boundary conditions, such as sea-ice and ice-sheet extent, are prescribed using reconstructed data records. In the more recent $2\times CO_2$ studies, boundary conditions are likely to be computed.

The features of the 18 ka BP climate listed in Table 8.1 are taken from the Cooperative Holocene Mapping Project (COHMAP) studies (see Kutzbach and Wright, 1985; Kutzbach and Guetter, 1986; COHMAP, 1988). COHMAP simulated global climate at 18, 15, 12, 9, 6, 3 and 0 ka BP using the Community Climate Model developed by the National Center for Atmospheric Research (NCAR).

The boundary conditions for these simulations are shown in Figure 3.1. Orbital forcing is represented by the seasonal distribution of Northern Hemisphere radiation. At 18 ka BP, the seasonal and latitudinal distributions of insolation are similar to those of today. (The radiation changes prior to 18 ka BP, however, were very different to those experienced prior to the present day. Hence climate conditions at 18 ka and 0 ka BP are very different, despite the similar pattern of insolation at these times.) After 18 ka BP, as the Earth–Sun distance shortened in the northern summer, and as obliquity increased, the seasonal radiation cycle of the Northern Hemisphere grew more pronounced, while that of the Southern Hemisphere weakened. In the Northern Hemisphere, maximum seasonality was reached at about 9 ka BP. At this time, Northern Hemisphere radiation, on average, was 8% greater in July than at the present day, and 8% less in January. Since 9 ka BP, the strength of the Northern Hemisphere seasonal radiation cycle has reduced, and that of the Southern Hemisphere increased. In addition to the boundary condition changes in insolation, changes in land ice, global mean SST, aerosol loading and atmospheric CO_2 are also incorporated in the COHMAP simulations (Figure 3.1). In the case of the Holocene thermal optimum simulation at 6 ka BP, only atmospheric CO_2 and the distribution of solar radiation differ from their present-day values.

The major features of the palaeoclimate data are reproduced reasonably well by the COHMAP model. The principal results from the Northern Hemisphere simulations for 18 ka BP and the present day are summarized in Figure 8.1 and in Table 8.1. Two major features of the Northern Hemisphere glacial circulation can be identified.

The first is the splitting of the westerly jet stream over the North American (Laurentide) ice sheet. At 18 ka BP the extensive and elevated (3 km thick) Laurentide ice sheet is responsible for splitting the flow of the jet stream in winter across North America. The upper branch follows the northern edge of the ice sheet and the southern branch is diverted south of the position of the present-day (single) jet over North America. Because of the steep temperature gradients at the southern margins of the Laurentide ice sheet and the extensive North Atlantic sea-ice field, the southern branch of the jet stream is strengthened across North America and the North Atlantic. Similarly, the flow of the westerly jet stream to the south of the European ice sheet (i.e. over the southern part of central and eastern Europe) is strengthened.

The second notable feature of the glacial simulation (Figure 8.1) is the formation of glacial anticyclones, due to higher pressure, over the Laurentide and Fennoscandinavian ice sheets. The dominance of easterly winds in the American northwest, and hence of colder and drier conditions, is supported by the palaeoclimate data (Kutzbach and Wright, 1985; COHMAP, 1988). In the American southwest, the southward displacement of the jet stream and storm tracks is associated with higher lake levels and the expansion of woodlands. Palaeoclimate reconstructions of conditions at 18 ka BP show that past effective moisture is greater in an area extending from California, across the Rockies and down into Texas (COHMAP, 1988).

Over the Northeast Atlantic and the British Isles, the combined effect of the changes in the jet stream and the presence of glacial anticyclones over the Laurentide and Fennoscandinavian ice sheets is to divert the westerly surface winds southwards of their present position. In addition, the strong glacial anticyclone over the Fennoscandinavian ice sheet causes southerly and easterly surface winds to be more common over Britain than at the present day. Westerly surface winds are, therefore, less common and the dominant air masses over the British Isles, and over northwest Europe in general, are continental rather than oceanic in nature. While Britain may have experienced limited westerly flow during the last glaciation, the colder

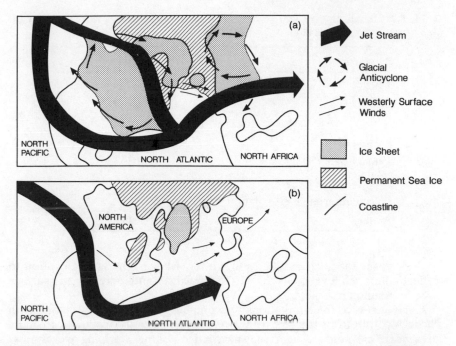

Figure 8.1 COHMAP GCM simulations for (a) 18 ka BP, and (b) the present day. (After COHMAP, 1988.)

SSTs and extensive sea ice in the North Atlantic at this time make the associated air masses colder and drier than at present.

The combined effect of all these changes is to make conditions at 18 ka BP in northwest Europe and the British Isles much colder and drier than at present, and to increase the strength of the seasonal cycle. This is demonstrated by the changes in temperature, precipitation and evaporation estimated by the NCAR GCM (Kutzbach and Wright, 1985). Estimates for Europe, expressed as departures from present-day values, are given in Table 8.2.

The COHMAP studies demonstrate the strong influence of the Laurentide ice sheet on the climate of the Northern Hemisphere in general, and on western Europe in particular. They also provide good examples of GCM snapshot simulations. We next consider the second type of GCM palaeoclimate simulation, sensitivity studies (Street-Perrott, 1991).

Sensitivity studies have been performed to assess the relative influence of changed boundary conditions on climate during a period of glaciation. In a study using the Goddard Institute for Space Studies GCM, for example, Rind (1987) assessed the individual influence and the combined effect on the climate system of lower North Atlantic SSTs; a 10 m thick ice sheet; and increased land elevation. Lower SSTs in the North Atlantic reduce latent heat fluxes and lead to drier conditions in Europe throughout the year. The presence of 10 m thick ice sheets over the land cools the surface in summer

Table 8.2 Climate change over Europe at 18 ka BP simulated by the NCAR GCM: departures from present day.

Parameter	Change
Temperature (°C)	
January	−20 to −25
July	0 to −5
Precipitation (%)	
January	−10 to −15
July	−5 to −10
Precipitation minus evaporation (%)	
Annual	−70 to −75

Source: Kutzbach and Wright, 1985

and increases the surface albedo in winter. Raising the land elevation to reflect falling sea level cools the surface in winter and increases the surface albedo in summer.

Comparison of the simulations for summer using all three factors, with those incorporating the first one or two factors only, suggests that the presence of elevated ice sheets may have led to summer warming over Eurasia at 18 ka BP (Rind, 1987). Rind demonstrates that the elevated model ice sheets induce atmospheric subsidence at their margins, so that cyclonic flow is established over parts of the continent. This in turn induces a flow of warm air from the south and also leads to increases in potential energy and eddy activity. The full simulation results reflect the balance of the conflicting influences from the three boundary condition changes (Rind, 1987).

The Laurentide ice sheet is known to have persisted into the warm Holocene period. The impact of this residual ice sheet on the climate system at 9 ka BP has been investigated in a study using the UK Meteorological Office GCM (Mitchell *et al.*, 1988). A two-centred ice dome with a maximum height of 2000 m, based on the reconstruction of Boulton *et al.* (1985), was incorporated into the GCM topography. Compared with model runs without a residual ice sheet, the simulation shows lower North Atlantic SSTs to the east of the ice sheet, thicker sea ice, and reduced warming over Europe and northern Asia. The warming over Europe is 1–2°C lower than in the simulation without a residual ice sheet (Mitchell *et al.*, 1988).

The sensitivity of regional climate to changes in the size and configuration of continental ice sheets is demonstrated by Shinn and Barron (1989). In this study, the NCAR Community Climate Model was used to simulate winter conditions at 18 ka BP. Two experiments were performed. In the minimum case, the ice-sheet limits were based on those of Boulton *et al.* (1985). In the British Isles an ice sheet just touches the north Wales coast but does not reach the Wash at a comparable latitude on the east coast. The total global ice volume implied by these boundary conditions is less than that indicated by marine oxygen isotope data. The maximum case is based on data from

Table 8.3 Major differences between CCM GCM simulations with Northern Hemisphere ice sheets at their maximum and minimum Last Glacial extents, as identified by Shinn and Barron (1989).

In the simulation with ice sheets at their maximum extent:

- Higher temperatures occur south and east of the ice sheets.

- A fourth geopotential height trough develops over the East Pacific. The amplitude of the planetary waves increases at all longitudes.

- Baroclinicity increases. Location of maximum jet stream speeds shifts polewards.

- Locations of maximum jet stream speeds and thermal gradients occur in the Atlantic, whereas they occur in the Pacific with ice sheets at their minimum extent.

- Split structure of the jet stream over North America is more defined.

- The cyclonic flow over eastern Siberia breaks down, probably due to a blocking effect of the larger Siberian ice sheet.

Hughes (1985) indicating extensive glaciation over Siberia and Alaska. The entire British Isles is covered by an ice sheet which extends across the English Channel to northeast Brittany.

Output from the two model runs is summarized in Table 8.3 (Shinn and Barron, 1989). When the maximum ice-sheet simulation is compared with the minimum simulation, higher temperatures are found over the North Atlantic, to the south of the Siberian ice sheet, in the Northwest Pacific and over central USA. At first sight, these results are surprising. It is suggested that they are related to a latitudinal shift in the planetary waves in the maximum case, allowing greater advection of warm air northwards and cold air southwards of a particular latitude (Shinn and Barron, 1989). This study agrees with the COHMAP studies in demonstrating that the behaviour of the westerly jet stream is related to the size of the Laurentide ice sheet, although it is difficult to separate physical from thermal influences. It is concluded that the splitting of the westerly jet stream over the maximum-size ice sheet is the result of the *physical* presence of the ice. In contrast, it is concluded that the latitudinal location of the jet stream cores over the North Pacific and North Atlantic oceans is related to the *thermal* influence.

Although the reliability of the model studies discussed above is uncertain, they demonstrate that the general circulation is very sensitive to the location, extent and altitude of ice sheets. So far as the climate of western Europe is concerned, it appears that glaciation over North America and Greenland may be more critical than ice build-up over Fennoscandinavia. These studies also provide information on global and large-scale regional changes in temperature at selected times in the past (see, for example, Table 8.2). GCMs cannot, however, be used to estimate climate conditions at specific locations (see Section 4.4). In order to assess past climate conditions at this scale of detail, it is necessary to look at palaeoclimate reconstructions based on proxy data.

8.3 Reconstructions based on palaeoclimate proxy data

It is not the aim of this section to describe the techniques of palaeoclimate reconstruction, or to assess the reliability of the various proxy indicators. Excellent detailed accounts are available elsewhere (Crowley, 1983; Bradley, 1985). Here we summarize a range of palaeoclimate studies describing climate conditions over northwest Europe, including our selected study areas in the British Isles. The focus is on the Last Glacial Maximum (about 18 ka BP), the Late-glacial and post-glacial periods including the Younger Dryas/Loch Lomond Advance (about 11 ka BP), and the Holocene (10 ka BP onwards). This evidence is presented in two sets of tables.

The first set, Tables 8.4 and 8.5, describes the general climatic conditions of the glacial and Holocene periods, including the prevailing features of the circulation system. The second set of tables provides quantitative estimates of temperature (Tables 8.6, 8.7 and 8.8) and precipitation (Table 8.9). The estimated temperature and the temperature change are shown for each study listed in Tables 8.6–8.8. Where the authors do not provide a figure for the change in temperature, we have estimated it from present-day records (our estimates are indicated by brackets in the final column of the tables). These tables are not intended to be exhaustive, but serve to demonstrate the range and quality of evidence which is likely to be available for any particular region.

The studies listed in these tables are based on a wide range of proxy indicators, including pollen sequences (for example, Pennington, 1970; Huntley and Birks, 1983; Huntley and Prentice, 1988; Guiot *et al.*, 1989; de Beaulieu *et al.*, 1991), fossil beetle remains (Coope, 1975; Atkinson *et al.*, 1987) and geomorphological features (for example, Bishop and Coope, 1977; Williams, 1975). Some studies are based on a synthesis of records (for example, Gray and Lowe, 1977; Lamb, 1977a; 1977b).

Estimates of the errors in the quantitative results are not given and are, in fact, largely unavailable. A particular problem is that, when interpreting many of the proxy indicators, it is difficult to distinguish between the effects of changes in precipitation, temperature, evaporation, drainage and non-climatic parameters. Therefore, these estimates should be treated with caution.

Given the uncertainty of the proxy data presented, and their limited geographical distribution, it is not possible to estimate parameters such as the average summer temperature at specific sites during the Last Glacial Maximum. We can, however, extract best estimates of the range of temperature change applicable to the general regions of northern England, Scotland and Wales. These are given in Table 8.10.

It is less easy to extract quantitative estimates of precipitation changes. As an approximate estimate based on the palaeoclimate data, we conclude that precipitation may have been 50% lower than today at the Last Glacial Maximum in the regions of interest. Precipitation has fluctuated throughout the post-glacial and Holocene but, in the British Isles at least, is unlikely to have been significantly higher than today at any time during these periods.

Table 8.4 General climate conditions and prevailing circulation patterns over northwest Europe in glacial periods. Studies are based on proxy palaeoclimate data and cover the Last Glacial Maximum (LGM), the Late-glacial including the Younger Dryas and Loch Lomond Advance (LLA), and the post-glacial period up to about 10 ka BP.

Reference	Climate characteristics
Pennington, 1970	NW Europe was warmer than the European continent during the late glacial.
Williams, 1975	LGM in British Isles: temperature gradients probably similar to those of today in winter (about 1.5°C per 100 km). Stronger seasonal cycle (about 30°C). Shorter, relatively warm summers.
CLIMAP, 1976	LGM : marked steepening of thermal gradients along the polar front system, particularly in the North Atlantic.
Gray and Lowe, 1977	Scotland, Late-glacial: dominance of S/SE winds bringing snow to C Grampians. Vigorous circulation with more frequent fronts and stronger winds than present.
Lamb, 1977a; 1977b	British Isles, LGM: highly meridional circulation. Possibly higher lapse rate. Greater year-to-year variability. Generally E surface winds in winter, lighter and more variable E winds in summer. Strong N winds all year.
Flohn, 1983	British Isles, LGM: frequent blocking anticyclones.
Huntley and Birks, 1983	NW Europe, Late-glacial: steeper climatic gradients than today.
Atkinson et al., 1987	C England, 14.5–13 ka BP: more continental, seasonal range of 30–35°C compared with 14°C today. 12.5 ka: also more continental, colder winters but annual mean temperature similar to today.
Lawson, 1987	LLA, Sutherland, Scotland: glacial accumulation aided by snow blown onto glaciers by SW and possibly NW winds, ie. principal snow bearing winds from S.
Keffer et al., 1988	North Atlantic, LGM: polar front has a more zonal alignment with steeper thermal gradient to S. Storm tracks shifted further S. W jet stream stronger than today.
Legrand et al., 1988b	LGM: increased wind speeds, more extensive arid areas and exposed continental shelves, more meridional circulation.
Dansgaard et al., 1989	Younger Dryas: North Atlantic colder, drier and stormier than present. S Greenland then warmed 7°C in 50 a, climate also milder, less stormy, more humid with weaker temperature/precipitation (N–S) gradients.
Tipping, 1989	13.5–10 ka BP: growth of trees on W Scottish seaboard suppressed by stronger, more westerly winds.

Table 8.5 General climate conditions and prevailing circulation patterns over northwest Europe in the Holocene. All the studies are based on proxy palaeoclimate data reconstructions.

Reference	Climate characteristics
Godwin, 1975	British Isles, about 8–6 ka BP: oceanicity pronounced in south and west, continentality in the east.
Lamb, 1977a; 1977b	Europe, general circulation reconstructions, Jan and July: Holocene dominated by westerly winds, year-to-year variability probably less than today. 8.5 ka BP: dry anticyclonic conditions, warmer summers. After this time, circulation weakened, more zonally (W–E) orientated with action centres located further N. 6.5 ka BP: opening out and eastward displacement of N American cold trough, surface westerlies spread further into Europe. 4.5 ka BP: westward shift of European surface pressure patterns. Very weak circulation related to colder winters. 2.5 ka BP: NW and C Europe more exposed to NW air, more penetration of westerlies, generally colder.
Keatinge and Dickson, 1979	Orkney, about 5 ka BP: evidence of increased wind speeds.
Huntley and Birks, 1983	Europe, some evidence for steepening of climate gradients (both N–S and E–W) following the thermal optimum.
Flohn and Fantechi, 1984	Europe, 8–6 ka BP: winter circulation very similar to present day.
Huntley and Prentice, 1988	6 ka BP: C and N Europe summers warmer than today. Winters warmer in N, but colder in NW, C and possibly S Europe. Reduced latitudinal gradient.
Tipping, 1989	Scotland, 10.3 ka BP: weakening of the atmospheric circulation system and increased wind transport from the S.

8.4 Discussion

The main features of the palaeoclimate data reconstructions summarized in Tables 8.4–8.10 show general agreement with the GCM palaeoclimate simulations discussed in Section 8.2.

During glacial episodes the flow of the westerly jet stream is split over the Laurentide ice sheet. The strength of the southern branch and of the flow across the North Atlantic into the southern part of central and eastern Europe is increased. The combined effect of the changes in the westerly jet stream and the establishment of glacial anticyclones over the Laurentide and Fennoscandinavian ice sheets is to push the major North Atlantic storm

Table 8.6 Estimates of European temperatures in the Last Glacial Maximum (about 18 ka BP). The departure from present-day temperature levels is indicated in the final column (figures in brackets indicate our own estimates based on the authors' data).

Reference	Region	Estimated temperature (°C)	Temperature change (°C)
Conolly and Dahl, 1970	Scotland	JJA: 25 (max)	JJA: max −4.5
	NW England	JJA: 28 (max)	JJA: max −5.6
	Wales	JJA: 28 (max)	JJA: max −5.2
Coope, 1975	C England	July: 8	(July: −8)
McIntyre et al., 1976	W Europe		Yr: −15/−18
Lamb, 1977a	British Isles	JA: 6	Yr: −10/−12
		DJF: −9	
	C Europe		Yr: −15/−16
Watson, 1977	Midlands and East Anglia	July: 5	(July: −11)
Sutherland et al., 1984	St Kilda, Scotland	July: 4	July: −9
Atkinson et al., 1987	British Isles	JJA: < 10	(JJA: > −5)
		DJF: −16 to −20/−25	(DJF: −26/−19)
West, 1988	C England		
Watson, 1977*		Yr: −5 to −10	Yr: −25
		July: 10–15	July: −5/−10
Williams, 1975*		Jan: −25	(Jan: −28)
		June: 10	(June: −4)
		Yr: −8/−10	(Yr: −20/−17)
Kolstrup, 1980*	Netherlands	Jan: −20	(Jan: −18)
		July: 5–10	(July: −12/−7)
Guiot et al., 1989	France, Grande Pile		Yr: −12 min
	Les Échets		Yr: −12
Rossignol−Strick and Planchais, 1989	S Italy	July: 24–21	July: −1/−4
		Jan: 0–14	Jan: −10/−24
Vandenberghe and Kasse, 1989 (early Pleistocene glaciation)	S Netherlands	Yr: −6 to −4	(Yr: −16/−13)
	N Belgium	JJA: < 10	(JJA: > −7)
de Beaulieu et al., 1991	France: Lac du Bouchet	Yr: −4	Yr: −11
		Jan: −18	Jan: −16
		July: 10	July: −6

* Data taken from West, 1988.

Table 8.7 Estimates of temperatures in Britain in the Younger Dryas (about 11 ka BP). The departure from present day temperature levels is indicated in the final column (figures in brackets indicate our own estimates based on the authors' data).

Reference	Region	Estimated temperature (°C)	Temperature change (°C)
Pennington, 1970	NW England	JJAS: 6.3 Jan: −7.5 Yr: < 0	(JJAS: −8) (Jan: −4) (Yr: > −9/−10)
Coope, 1975	C England	July: 10	(July: −6)
Bishop and Coope, 1977 Kerney et al., 1964*	SW Scotland N England Kent	July: 8/9 Yr: 3 Jan: −6 July: 12/13	July: −2/+3 Yr: −4/−5 (Yr: −7) (Jan: −1/−2) (July: −5/−4)
Gray and Lowe, 1977 Péwé, 1966**	Scotland Mull and W Ross	Yr: at least −1, possibly up to −6/−8	(Yr: at least −9, possibly up to −15/−17)
Sissons, 1974**	Mull Grampians	Jan: at least −7, possibly up to −17/−19 July: 5 July: 10	(Jan: at least −12, possibly up to −22/−24) (July: −13) July: −9
Gray, 1982	Snowdonia, north Wales	July: ~8.5	
Atkinson et al., 1987	British Isles	Yr: −2 to −5	(Yr: −15/−11)

* Data taken from Bishop and Coope, 1977.
** Data taken from Gray and Lowe, 1977.

tracks southwards of their present position. In addition, surface westerly winds are less dominant over the British Isles, while easterly and southerly winds are more common than at the present day (Gray and Lowe, 1977; Lamb, 1977a; 1977b; Lawson, 1987). The southerly winds have a greater moisture capacity than the easterlies, and so are thought to play a more important role in ice sheet and glacier accumulation. There is some evidence that southerly winds are more frequent over Scotland than over north Wales and the Lake District (Gray, 1982). In glacial episodes, therefore, the climate of the British Isles is more continental than at the present day, with an enhanced seasonal cycle and stronger thermal gradients (both north–south and east–west). Across northwestern Europe, conditions are generally colder, drier and stormier. Similar, but weaker, climate changes are seen during stadial episodes such as the Younger Dryas.

Conditions at the peak of interglacial episodes are, of course, quite similar to those of the present day. For example, westerly winds tend to prevail over

Table 8.8 Estimates of European temperatures in interglacials and interstadials, including the Eemian (about 125 ka BP) and the Holocene (about 10 ka BP). The departure from present-day temperature levels is indicated in the final column (figures in brackets indicate our own estimates based on the authors' data).

Reference	Region: period	Estimated temperature (°C)	Temperature change (°C)
Pennington, 1970	NW England: Alleröd	July > 10	
Coope, 1975	C England: Ipswichian Upton Warren	July: 20 July: 17.5	(July: +4) (July: +1.5)
Bishop and Coope, 1977	SW Scotland: 13 ka BP 9640±180 a BP	July: 15	(July: +1) July: warmer
Lamb, 1977a	British Isles: 6 ka BP	JA: 18 DJF: 5 Yr: 10.7	JA: +2 DJF: +1 Yr: +1.6
Flohn and Fantechi, 1984	Europe: Eemian Holocene		Yr: +2/2.5 Yr: +1.5
Atkinson et al., 1987	British Isles: 12.5 ka BP	Summer max: 17 Winter min: 0–1	Yr: little change
Huntley and Prentice, 1988	Scotland, W coast, islands and borders: 6 ka BP		July: 0 to +2 July: +2 to +4
Guiot et al., 1989	France, Grande Pile and Les Échets: Eemian		Yr: +2
de Beaulieu et al., 1991	France, Lac du Bouchet: Eemian	Yr: 10	Yr: +3

the British Isles. Nevertheless, differences do exist. In addition to the general warmth, conditions in the British Isles are more oceanic (i.e. the seasonal cycle is slightly weaker). In comparison with the present day, the strength of the zonal (west–east) circulation is somewhat reduced. Thermal gradients are also weaker.

Table 8.9 Estimates of European precipitation in the last glacial–interglacial cycle including the Last Glacial Maximum (LGM) about 18 ka BP and the Holocene (the last 10 ka).

Reference	Region: period	Precipitation regime
Manley, 1975	British Isles uplands: prior to LGM	Possibly 150% of present-day prec.
Williams, 1975	LGM: East Anglia Midlands Scotland	< 50% present prec. < 33% present prec. 60% present prec. All regions: light winter snowfall
Gray and Lowe, 1977	Scotland: post-glacial	Probably a change in distribution rather than in total rainfall
Lamb, 1977a	British Isles: Holocene	Prec. up to 15% > than today, some drier periods
Keatinge and Dickson, 1979	Orkney: 3.4 ka BP onwards	Bog development due to increased P:E ratios, increased oceanicity and grazing pressure
Flohn, 1983	Europe: LGM	Prec. and evaporation reduced by 20–25%
Huntley and Prentice, 1988	Europe: 6 ka BP	Less reliable snow in N, prec. increase in E, shorter period of Mediterranean drought
Guiot *et al.*, 1989	France: last 140 ka	Less precip. in glacial and more in warm periods Les Échets: today 800 mm; range over past 140 ka: 100 to 900 mm Grande Pile: today 1080 mm; range over past 140 ka: 280–1580 mm
Rossignol-Strick and Planchais, 1989	S Italy: LGM	Prec. reduced by 440–540 mm, to 450–350 mm
de Beaulieu *et al.*, 1991	France, Lac du Bouchet: last 140 ka	LGM: Yr: 300 mm July: 5 mm, Jan: 25 mm Interglacial max: Yr: 960 mm July and Jan: 100 mm

Table 8.10 Best estimates of the range of temperature changes for northwest England, Scotland and Wales in selected past episodes, expressed as departures from present-day conditions.

Episode	Temperature change
Last Glacial Maximum (18 ka BP)	Mean annual temperature 10–20°C lower, i.e. below freezing.
	Cooling greatest in winter (–20°C to –25°C).
	Summer cooling considerably less (–5°C to –10°C).
Younger Dryas (11 ka BP)	Mean annual temperature about 10°C lower, i.e. close to, probably just below, freezing.
	Seasonal change is uncertain, but cooling probably weakest in winter.
Interglacials (Ipswichian and Holocene) (125 and 10 ka BP)	Maximum annual warming of 2°C.
	Warming in summer probably slightly greater than in winter (e.g., 2°C and 1°C, respectively).

Given all the uncertainties, it is encouraging that estimates of the magnitude of temperature changes for the Last Glacial Maximum and the Holocene thermal optimum taken from GCMs and from palaeoclimate reconstructions show general agreement (see Tables 8.2 and 8.10). This suggests that GCMs can be used to infer circulation changes in the past, and to simulate future climate, as described in the next chapter. However, because the model boundary conditions, such as ice-sheet extent and sea-surface temperatures, are prescribed from the palaeoclimate data, GCMs cannot be used to check the palaeoclimate reconstructions of boundary conditions and, similarly, the model results cannot be validated using palaeoclimate data.

9 Models of future climate

9.1 Introduction

The use of general circulation models (GCMs) in palaeoclimate reconstruction is described in the previous chapter. Estimates of past climate conditions in the British Isles and our selected study regions (Caithness, Cumbria and Wales) are also discussed. Here, the use of GCMs as a guide to the future climate conditions which may be experienced in northwestern Europe in a potential greenhouse gas-induced warming episode is described. In the final part of this chapter, long-term simulations of future climate produced by orbitally forced climate models are presented.

9.2 Greenhouse gas-induced warming

9.2.1 Evaluation of grid-point GCM data for northwestern Europe

Despite their many uncertainties and known errors, GCMs offer the greatest potential for developing regional scenarios of climate changes associated with enhanced greenhouse warming. Increasingly sophisticated methodologies are being developed to overcome the problems caused by the low spatial resolution of GCM grid-point output (see Chapter 4 and, for example, Kim *et al.*, 1984; Wilks, 1989; Karl *et al.*, 1990; Wigley *et al.*, 1990). A limited number of high-resolution models have been run (Cubasch and Cess, 1990; Mitchell *et al.*, 1990), but output from these simulations is not yet widely available for use in impact assessment studies.

In this chapter we use data for northwestern Europe from five GCMs to demonstrate some of the problems encountered when trying to use the information from individual model grid-points in climate impact studies. The analysis is restricted to equilibrium experiments because of the greater experience with these models, and because of the additional problems which arise when models are run in transient mode.

Seasonal temperature and precipitation data, expressed as differences between the $2 \times CO_2$ (perturbed) and $1 \times CO_2$ (control) runs, have been obtained for the five GCMs shown in Table 4.1. These are research data only and we have no guarantee of their reliability, other than through comparison with published GCM studies. Although these data do not come from the most up-to-date model runs they are adequate for our purpose here, which is to demonstrate some of the problems which arise when using GCM output for regional climate impact studies.

The locations of the grid-points are shown in Figure 9.1. In this assessment, the definition of the northwestern European area is constrained by the GCM grid scales and is taken to be 50–67.5°N × 11.25°W–10°E. The

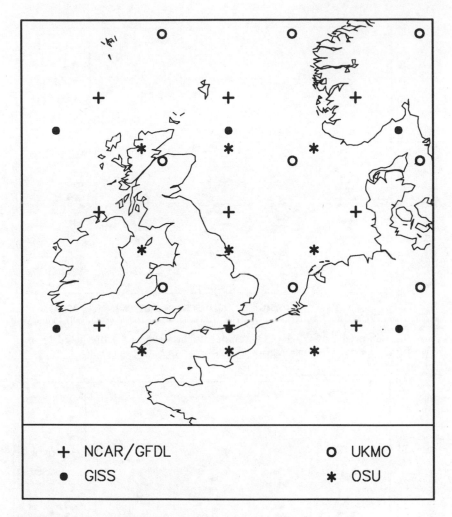

Figure 9.1 Location of GCM grid-points over western Europe.
UKMO: UK Meteorological Office
NCAR: National Center for Atmospheric Research
GFDL: Geophysical Fluid Dynamics Laboratory
GISS: Goddard Institute for Space Studies
OSU: Oregon State University

GISS model has six grid-points located in this area, whereas the other models all have nine. The NCAR and GFDL models share the same grid.

The $2 \times CO_2 - 1 \times CO_2$ run differences for each model and for each season are shown in Figures 9.2 (temperature) and 9.3 (precipitation). In these composite diagrams, the GFDL model, for example, has nine values plotted for each season, corresponding with the nine grid-points in Figure 9.1.

For temperature, it can be seen that the greatest increases are shown by

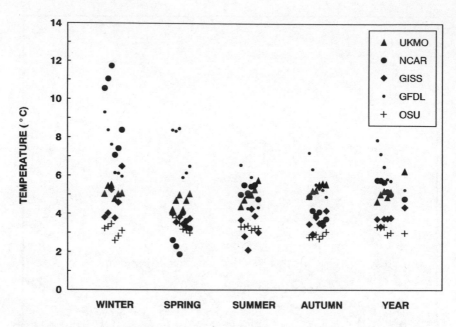

Figure 9.2 Seasonal $2\times CO_2 - 1\times CO_2$ temperature change (°C) predicted by five GCMs for the western European grid-points shown in Figure 9.1.

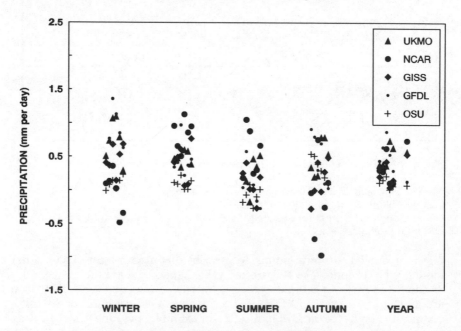

Figure 9.3 Seasonal $2\times CO_2 - 1\times CO_2$ precipitation change (mm per day) predicted by five GCMs for the western European grid-points shown in Figure 9.1.

the GFDL and NCAR models, particularly in winter (Figure 9.2). For the whole year, the increases are between 2.6°C and 7.9°C. The smallest temperature increases occur in the OSU GCM, with changes between 2.0°C and 4.0°C averaged over the whole year.

We might expect that the greatest changes in temperature, and possibly precipitation, would occur in the model with the greatest global sensitivity, i.e. the model with the largest $2\times CO_2-1\times CO_2$ global temperature change. Figure 4.3 indicates that, of these five particular GCMs, the OSU model is the least sensitive, and the UKMO model the most sensitive. We therefore expect that the greatest warming over northwestern Europe will be predicted by the UKMO GCM. This is not the case. There does not appear to be any consistent relationship between predicted $2\times CO_2$ temperature change over northwestern Europe and global $2\times CO_2$ temperature change (Figure 9.4).

In comparison with the predictions for temperature, the predictions for precipitation are much less consistent (Figure 9.3). All the models show lower rainfall in summer and autumn at some of the grid-points, generally those located in the south or east of the region. The NCAR model shows a decrease in rainfall of as much as 1 mm per day in autumn at the grid-point located close to the western Danish coast.

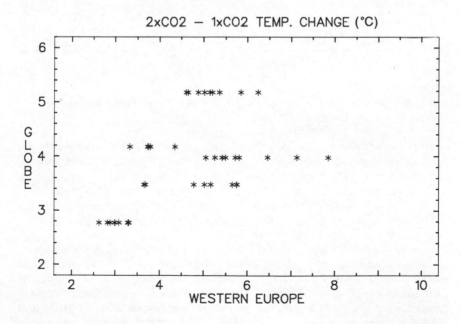

Figure 9.4 Global $2\times CO_2-1\times CO_2$ temperature change relative to western European $2\times CO_2-1\times CO_2$ temperature change (°C) for the five GCMs.

In Figures 9.5 and 9.6, we show seasonal profiles at 55.55°N for three grid-points of the GFDL and NCAR models (7.5°W, 0° and 7.5°E). The magnitude of the temperature changes predicted by the GFDL model is much greater than indicated by the NCAR model in spring and autumn (Figure 9.5). There are also some interesting differences in the seasonal distribution of change in the two models. The NCAR model suggests maximum changes in winter and summer, whereas the warming in each season is similar in the GFDL model.

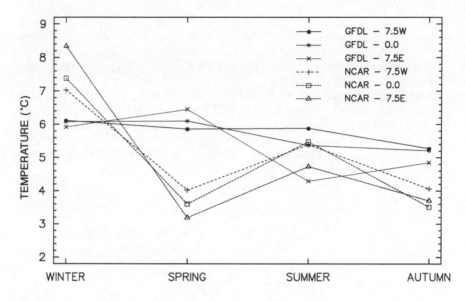

Figure 9.5 Seasonal variations in $2\times CO_2 - 1\times CO_2$ temperature change (°C) at specific grid-points for the GFDL and NCAR GCMs.

In the case of precipitation, the changes shown by the two models are entirely different (Figure 9.6). The NCAR model predicts a decrease in precipitation in winter and, at two grid-points, in the autumn. The maximum increase is indicated in spring. The GFDL model predicts a small decrease in summer at two grid-points, and an increase in the other seasons, peaking in the winter. The impacts of the climate changes suggested by each of these two models are likely to be very different.

Before the results from any GCM can be used in climate impact studies, its ability to simulate present-day climate should be assessed. As an example of the problems which may be encountered, we consider the ability of one particular version of one GCM to simulate observed seasonal variations of temperature over Britain (Figure 9.7). Here we compare GCM grid-point output with station data – a procedure not recommended for the purpose of model validation, but legitimate when the general suitability of GCMs for regional scenario construction is being assessed. The GCM control run data are from a version of the UKMO GCM and are compared with instrumental

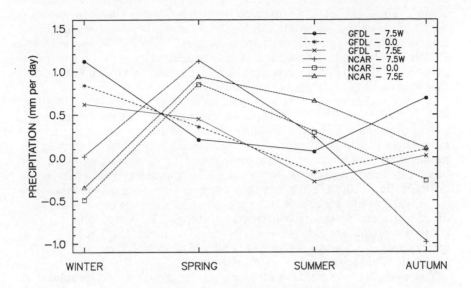

Figure 9.6 Seasonal variations in $2 \times CO_2 - 1 \times CO_2$ precipitation change (mm per day) at specific grid-points for the GFDL and NCAR GCMs.

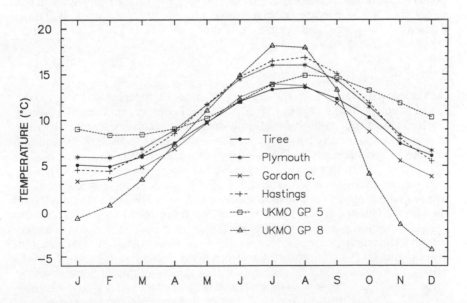

Figure 9.7 Comparison of present-day mean monthly temperature data (°C) from four UK station records with control-run output from the UKMO GCM (grid-point 5: 57.5°N 3.75°E, grid-point 8: 52.5°N 3.75°E).

193

data from four stations representing the north–south/east–west range of climate. These are Tiree in northwest Scotland, Gordon Castle in northeast Scotland, Hastings in southeast England, and Plymouth in southwest England. The chosen grid-points are located at 57.5°N 3.75°E and at 52.5°N 3.75°E. It should be stressed that this model version is typical of its generation. Its performance is considered to be no worse, and no better, than that of other contemporary models.

Neither grid-point provides a reliable estimate of present-day conditions in the British Isles. This is largely due to the unrealistic seasonal cycles at the two grid-points which can, in turn, be attributed to the fact that the northernmost grid-point represents a grid square which is treated by the model as an ocean box, while the southernmost square is treated as a land box. The seasonal cycle of temperature is, therefore, too weak in the northern box and too strong in the southern box. Clearly the simplified model geography cannot match the real-world complexity of an island lying on the western fringe of continental Europe. The geography of high-resolution GCMs should, however, be more realistic. The UKMO GCM, for example, has recently been run at a resolution of 2.5° latitude by 3.75° longitude (Cubasch and Cess, 1990; Mitchell *et al.*, 1990).

The examples presented so far demonstrate that it would be unwise to produce a regional estimate of change simply by taking model output from the grid-points located closest to the region of interest. A number of questions must be addressed. Which model, for example, is most reliable? Which grid-points are most representative of the region of interest? Can sub-grid-scale values be interpolated? There are no easy answers to any of these questions, but in Section 9.2.2 we discuss appropriate methodologies for investigating such issues.

9.2.2 Regional scenario development

At the present time, there are no *a priori* reasons to assume that any one GCM is better than any other. It is possible, however, that comparison of control run output and observational data may identify one model which performs best over a particular region. As numerical techniques for spatial validation become more widely available (Santer and Wigley, 1990; Wigley and Santer, 1990), this approach may become more practical.

If we consider the results from all available models, then at least we are aware of the range of variations in model estimates. The International Panel on Climate Change (IPCC) recently produced best estimates of changes in mean surface temperature, precipitation and soil moisture over five selected regions (Mitchell *et al.*, 1990, Table 5.1) from three high-resolution equilibrium GCMs. These are versions of the Canadian Climate Center, GFDL and UKMO GCMs. As explained at the beginning of Section 4.4.6, these results were adjusted to represent the expected transient global warming by 2030 assuming the Business-as-Usual (BaU) scenario for greenhouse gas emissions. Over southern Europe in summer the mean temperature is estimated to rise by 2–3°C, precipitation to decrease by 5–15% and soil moisture to decrease by 15–25% (Table 9.1). It should be noted that these results apply only to each region as a whole and cannot be applied to any one part individually.

Table 9.1 Estimates of changes in areal means of surface temperature, precipitation and soil moisture over southern Europe (35–50°N, 10°W–45°E), from pre-industrial times to 2030, assuming the IPCC Business-as-Usual scenario. Estimates are from three high-resolution GCMs: CCC: Canadian Climate Center; GFDL: Geophysical Fluid Dynamics Laboratory; UKMO: UK Meteorological Office.

Model	Temp. (°C) DJF	Temp. (°C) JJA	Prec. (%) DJF	Prec. (%) JJA	Soil Moisture (%) DJF	Soil Moisture (%) JJA
CCC	2	2	5	–15	0	–15
GFDL	2	2	10	–5	5	–15
UKMO	2	3	0	–15	–5	–25

Source: Mitchell *et al.*, 1990

In order to gauge the reliability of these regional estimates, the IPCC expert group assumed that the range of best estimates of global mean warming for an equivalent CO_2 doubling (i.e. 1.5–4.5°C) also represents the range of uncertainty associated with regional estimates of all climate variables (Mitchell *et al.*, 1990). The 1.5–4.5°C range implies that the regional estimates may actually lie within 70–145% of the given values. The group, therefore, concludes that 'confidence in these estimates is low' (Mitchell *et al.*, 1990).

More sophisticated methods are being developed for synthesizing grid-point results from GCMs into regional climate change scenarios. Wigley *et al.* (1992), for example, have developed a statistical method for averaging standardized patterns of temperature change from five different models to produce estimates which are not biased by the individual model sensitivities. This approach alone is not appropriate for precipitation because of the poor agreement between the different models, particularly in the tropics (30°S to 30°N). For precipitation, therefore, both the probability of a change in precipitation in a specified direction and the percentage change are estimated.

The methods of scenario development described above are all constrained by the spatial resolution and the grid-point mesh of each GCM. We found, for example, that the control run data from the two grid-points presented in Figure 9.6 were unrepresentative of the geography and climate of the British Isles. In this particular case it would be possible to interpolate between the two grid-points (one land-based and one ocean-based) to obtain a more realistic estimate of conditions in the British Isles. Given the location of the islands on the west coast of a continent, such an interpolation has some physical justification. This general concept has been employed in the development of statistical techniques for the interpolation of sub-grid-scale values from GCM grid-point data (see Section 4.4.2).

So far we have restricted our discussion to the variables most commonly recorded during GCM experiments: mean annual, seasonal and monthly temperature, precipitation and soil moisture. Estimates of other parameters can sometimes be obtained from modelling groups. If results are unobtainable, or are considered unreliable, then it may be possible to calculate values for the required variable from the available output. A numerical boundary layer model, for example, has been used to estimate regional-scale surface wind speed variations in the British Isles from grid-point vector mean wind speeds predicted by a GCM (Palutikof *et al.*, 1990). Changes in the mean state of a parameter may not, however, provide the most valuable information for input to climate change impact assessments (Robinson and Finkelstein, 1991). Information about regional changes in climate variability, including changes in the magnitude and frequency of extreme events, may be needed. The daily variability of temperature and precipitation is not accurately reproduced in the current generation of GCMs (Mearns *et al.*, 1990). Climatologists are, therefore, developing methodologies for the extrapolation of information about short-term (i.e. daily) events from GCMs using statistical models developed from the observed record (Gregory *et al.*, 1990; Chen and Robinson, 1991; Robinson, 1991).

This brief discussion of regional scenario construction demonstrates some recent advances in the development of methodologies for using GCM data in regional impact assessment. At the present time, however, all of these methodologies are constrained by the reliability of the current generation of equilibrium GCMs. Without major improvements in modelling, these methodologies can only be used to explore the *sensitivity* to climate change of, for example, hydrological and ecological systems. Reliable, quantitative regional predictions are still beyond our grasp.

If reliable regional scenarios could be constructed, over what time scale could they be applied? Assuming the IPCC BaU greenhouse gas-emission scenario, it is likely that equivalent CO_2 concentrations will double within 30–40 a (Houghton *et al.*, 1990). If stringent emission controls ·are introduced, then doubling may not occur until the end of the twenty-first century. Whatever the eventual doubling time, the full equilibrium warming will be delayed by a matter of some decades. The nature of this transient response is discussed in Section 4.4.6. In order to consider future conditions over longer time scales, it is necessary to consider the influence of orbital forcing.

9.3 Orbital forcing

Although climate models have been widely used to explore the reality of orbital–climate relationships and the potential mechanisms linking cause and effect (see Chapter 2), comparatively few model studies have attempted to predict future climate on the basis of orbital variations. We have, however, identified seven statistical/dynamical models and time-dependent ice-sheet models (see Tables 2.2 and 2.3) which do attempt such a prediction.

The principal characteristics of the seven models (Kukla and Kukla, 1972; Calder, 1974; Weertman, 1976; Imbrie and Imbrie, 1980; Berger *et al.*, 1981; Kukla *et al.*, 1981; Oerlemans and Van der Veen, 1984), together with details of model calibration and reliability, are presented in Table 9.2. Climate reconstructions for the last 120 ka and projections for the next 120 ka are shown in Figure 9.8. The total period for which output is available from each model is given in Table 9.2. With the exception of Study 6 (Kukla *et al.*, 1981), which calculates values of an astronomical climate index (referred to as ACLIN) for the next 1 Ma, none of the studies makes projections beyond 150 ka AP. Each study presents the climate output in a different way but, for comparative purposes, the curves shown in Figure 9.8 can be considered as climate indexes indicating relative warming and cooling. The index levels for the Last Glacial Maximum and for the Holocene thermal optimum are shown by the upper and lower horizontal lines, so that relative amplitudes can be compared.

Details of model calibration are given in Table 9.2. Results from all seven studies indicate at least general agreement with the geological record. However, spectral analyses of model results and marine $\delta^{18}O$ records have revealed shortcomings. Models 2 and 4 (Calder, 1974; Imbrie and Imbrie, 1980) exaggerate the amount of power at the precessional periodicities of 23 ka and 19 ka compared with the 41 ka obliquity periodicity. Neither of these two models, nor Model 6, produces sufficient power at the 100 ka eccentricity period (Imbrie and Imbrie, 1980; Kukla *et al.*, 1981). None of the models reflect the observed changes in the sensitivity of the climate system to the various types of orbital forcing (see Chapter 2). Other shortcomings can be recognized in the models, such as excessive sensitivity of output to changes in the values of model parameters.

Despite the shortcomings outlined here and in Section 2.3, we consider that the models provide a reasonably realistic guide to the range of future climate change caused by periodicities in the orbital parameters.

The model results of Figure 9.8 present a broadly consistent picture of climatic changes expected over the next 100 ka. The long-term cooling trend which began roughly 5 ka BP, following the Holocene warm period, is projected to continue throughout. This trend should, however, be punctuated by a number of warming episodes or, at the least, periods of slight amelioration. Models 5 and 6, for example, show a relatively warm peak about 30 ka AP, while Models 2–4 merely indicate a levelling off or a limited warming at this time.

Studies 1, 5 and 6 indicate a cold trough about 5 ka AP and most studies predict cold troughs at about 23 ka and 60 ka AP, the latter being particularly severe. Although the timing of these minima appears consistent, there are some differences between models in the relative amplitudes of the 'climate' curves, both in the representation of past events and the predictions for the future. The only future event which reaches or exceeds the severity of the Last Glacial Maximum in all studies is the glacial event projected for around 60 ka AP.

The warmth of the Holocene climate optimum is not expected to be exceeded until 120 ka AP (Study 6). The projected course of events between the suggested glacial maximum at 60 ka AP and the thermal optimum at 120 ka AP is somewhat uncertain. A warmer period is indicated about 75 ka AP (Studies 2, 3, 4, 6 and 7). This appears to be followed by a further glacial

Table 9.2 Description of models of future climate based on orbital changes.

STUDY 1	Kukla and Kukla, 1972
Orbital input	Winter insolation, 25–75°N (Vernekar, 1968)
Description	Rates of insolation change are calculated and positive/negative insolation regimes, identified.
Output	Annual rate of change in winter insolation, 150 ka BP–30 ka AP.
Calibration	3 ocean core records, coral reefs, continental records from N America and Europe. 'High degree of correlation exists' between identified insolation regimes and climate.
Reliability	Only encompasses 20 ka cycle: cannot provide complete explanation of 100 ka and longer cycles.
STUDY 2	Calder, 1974
Orbital input	Summer insolation 50°N, Vernekar (1972)
Description	Assumes ice volume grows in proportion to deficit of radiation below a certain point and melts in a different proportion above that point.
Output	Global ice volume, arbitrary units, 860 ka BP–120 ka AP.
Calibration	Core V28–238. Major variations reproduced and core stages identifiable but discrepancies exist, e.g. model is too warm at 170 ka BP and too cold in core stages 11 & 21.
Reliability	Very sensitive to small changes in parameter values. Spectral analysis: produces more power at 100 ka and less at 413 ka than Study 4, but at expense of spurious 200 ka power. Exaggerates power at 23 ka and 19 ka compared with 41 ka.
STUDY 3	Weertman, 1976
Orbital input	50°N radiation, averaged over high-radiation half-year (Vernekar, 1972).
Description	2-D ice sheet on flat land surface. Assumes ice behaves as perfectly plastic solid and uses glacier mechanics theory. Assumes ice sheet is only affected by changes in snowline elevation caused by radiation changes.
Output	Ice-sheet half-width (km), 300 ka BP–80 ka AP.
Calibration	Peaks and troughs of past 130 ka 'correlate well' with Shackleton and Opdyke's (1973) sea-level data.
Reliability	Main criticism is use of 'rather large, but not impossibly large, accumulation rate as well as ablation rate'. Very sensitive to changes in these rates and their ratio. Spectral analysis: power at 23 ka and 19 ka exaggerated, compared with 41 ka. Power also seen at 100 and 400 ka.
STUDY 4	Imbrie and Imbrie, 1980
Orbital input	July 65°N insolation (Berger, 1977; 1978a)
Description	Simple non-linear model simulates lag between orbital variations and ice-sheet response using 4 adjustable parameters to tune model to geological record of last 150 ka.
Output	Ice volume, arbitrary units, 500 ka BP–100 ka AP.

Table 9.2 Description of models of future climate based on orbital changes. (Continued)

Calibration	Same records as for tuning: ocean cores RC11-120, E49-18 and V28-238. Age of 6 specific events agrees well with radiometric dates and 'relative magnitude of all major features' reproduced 'satisfactorily' to 150 ka BP. Significant correlations back to 350 ka BP, then diminish (about time of last 413 ka eccentricity minimum).
Reliability	Spectral analysis: insufficient 100 ka power, too much at 23, 19 and 413 ka. All problems related to 100 ka failure.
STUDY 5	Berger *et al.*, 1981
Orbital input	Monthly insolation at 6 different months/latitudes.
Description	Autoregressive multivariate spectral model. Climate is function of insolation and climate of previous 3 ka.
Output	$\delta^{18}O$ (%), 400 ka BP–60 ka AP.
Calibration	0–468 ka BP, ^{18}O record of Hays *et al.*, 1976, concentrating on last 150 ka.
Reliability	Explains 65% of variance 15–60 ka BP; 55% 0–15 ka BP and 60–135 ka BP; 50% past 470 ka. Quasi-periods have correct relative amplitudes.
STUDY 6	Kukla *et al.*, 1981
Orbital input	3 orbital parameters, revision of Berger, 1978b.
Description	Combines 3 orbital variables in time-lag bivariant model. Empirical formula and improved formula describe link between orbital perturbation and insolation.
Output	ACLIN1 Climate index, 1 Ma BP–1 Ma AP.
Calibration	Ocean core V28-238 and other records (eg. coral terraces, pollen cores) for last 130 ka. Predicts interglacials over past 350 ka within dating accuracy. Relative amplitude of interstadials/cold minima of last 130 ka also predicted. Relative intensity of earlier cold events not reproduced well.
Reliability	Spectral analysis: too much power at 19, 23 and 400 ka, too little at 100 ka, but 400 ka peak is lower and more realistic than Model 4. 100 ka peak also more realistic. Modulating effect of eccentricity is not properly accounted for.
STUDY 7	Oerlemans and Van der Veen, 1984/Oerlemans, 1982
Orbital input	Summer insolation 65°N (Berger, 1978a)
Description	Numerical ice-sheet model with bedrock sinking, forced by moving snowline in response to insolation changes; mass-balance and ice-surface temperature are affected by forcing.
Output	Ice volume, 0–150 ka AP and 0–800 ka BP (Oerlemans, 1982).
Calibration	$\delta^{18}O$ records (e.g., Shackleton and Opdyke 1973; 1976). Records and model 'qualitatively similar'.
Reliability	Very sensitive to ice-accumulation rate and other parameters, so fine tuning is impossible. 'Predictability' is considered low because of non-linear instabilities.

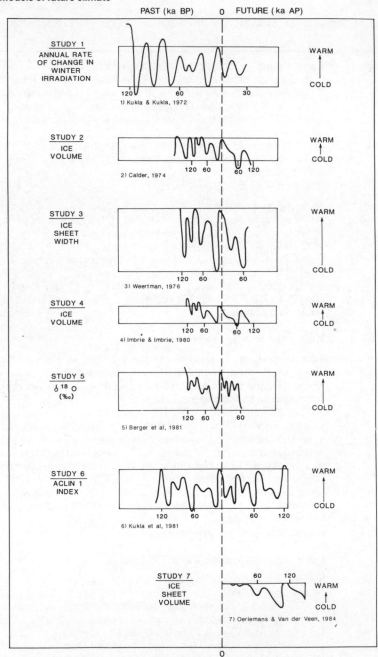

Figure 9.8 Predictions of future climate from seven orbital change models. To indicate the relative scale, the lower and upper horizontal lines on each study represent index values for, respectively, the Last Glacial Maximum and the Holocene climate optimum. All dates are given in ka. (From Goodess *et al.*, 1988.)

event about 100 ka AP (Studies 2, 4, 6 and 7), although models differ as to whether this event is more or less severe than that at about 60 ka AP.

All seven studies indicate the persistence of glacial–interglacial cycles typical of the Quaternary. The sawtooth character of these cycles is also evident in the model results, i.e. rapid deglaciation and warming followed by a slower, often interrupted, cooling towards the next glaciation. Unfortunately, only Study 6 gives projections beyond the next interglacial peak at 120 ka AP. The ACLIN1 index predicts a further eight peaks equal to or exceeding that of the Holocene optimum, at about 170 ka, 240 ka, 280 ka, 620 ka, 720 ka, 910 ka, 980 ka and 1 Ma AP. Seven ACLIN1 troughs equal to or exceeding that of the Last Glacial Maximum are predicted at about 60 ka, 380 ka, 420 ka, 670 ka, 800 ka, 890 ka and 960 ka AP. There are a number of other minima which approach these levels such as, over the next 300 ka, those at about 23 ka, 110 ka, 130 ka, 180 ka and 230 ka AP. The number of both warm and cold ACLIN1 extremes calculated for the last 1 Ma is virtually the same as predicted over the next 1 Ma. Furthermore, the range of index values (about 0.5 to 5) is the same for both the past and the next 1 Ma periods. This suggests that extreme conditions likely to be experienced over the next 1 Ma will not exceed the extremes of the last 1 Ma.

Although there are some discrepancies between the future sequences of climate predicted by the models, the major changes are common to all seven studies. To summarize, these models suggest that, if we ignore the potential for anthropogenic change, the world's climate should be just beginning a slow deterioration towards glacial conditions. Oscillatory cooling is expected to continue with progressively colder episodes at about 5 ka, 23 ka and 60 ka AP. The final extreme around 60 ka AP is expected to rival the intensity of the Last Glacial Maximum and should be followed by a gradual shift towards warmer conditions. The models indicate that climates as warm as the present day are relatively rare. The world is not expected to return to conditions matching the Holocene thermal optimum until about 120 ka AP.

In view of the criticisms which can be directed at these studies, it is encouraging that preliminary results from a new 2.5-dimensional physical climate model indicate a glacial maximum in the Northern Hemisphere beginning about 55 ka AP in the absence of enhanced greenhouse warming (Berger *et al.*, 1991b). This model is considerably more realistic than the simple numerical models discussed above. The principal features of the model are outlined in Section 2.3. Simulations for the next 80 ka AP are shown in Figures 2.7 and 9.9 (Tricot *et al.*, 1989; Berger *et al.*, 1991b).

The model studies described so far consider only the response of the climate system to orbital forcing. The influence of shorter-term forcing factors, including anthropogenic effects (see Chapters 3 and 4), is ignored. Thus, even the most realistic of the models has its limitations as a guide to future climate. Berger's model has, however, been used to test the sensitivity of orbital forcing to anthropogenic greenhouse gas-induced warming (Berger *et al.*, 1991b). In this sensitivity study, described in Chapter 5, it is assumed that enhanced greenhouse warming will melt the Greenland ice sheet totally. Thus, the model simulation of the next 80 ka is

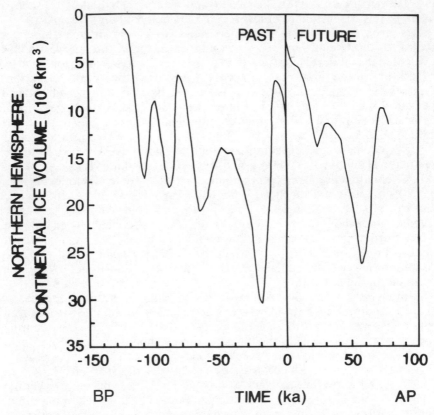

Figure 9.9 Model simulation of continental ice volume for the last 120 ka and the next 80 ka. The model is a 2.5-dimensional model, forced by orbital changes. (From Berger *et al.*, 1991b.)

repeated with no Greenland ice sheet. The main results of this sensitivity study are that ice sheets do not reappear in the Northern Hemisphere until about 15 ka AP, and that the next glaciation is delayed by about 5 ka to around 60 ka AP and is less extensive (Figure 5.1; Berger *et al.*, 1990).

The most recent modelling results presented here confirm that, potential anthropogenic forcing apart, the pattern and range of climatic conditions likely to be experienced over the next 1 Ma will be close to those experienced over the last 1 Ma. This implies that it is reasonable to use the reconstructed record of Quaternary climate as a guide to future conditions. These studies also provide some indication of the ways in which future climate sequences may diverge from past sequences because of enhanced greenhouse warming.

10 The uncertainties in future climate assessment

10.1 The end of the Holocene: what next?

The present interglacial (the Holocene) began about 10 ka BP. An analysis of the length of previous interglacials, defined here as periods of Temperate climate conditions in the British Isles, has been carried out by the authors in a study of the implications of long-term climate change for radioactive waste disposal (Goodess et al., 1992). The analysis was based on the SPECMAP climate index (Imbrie et al., 1984) and on regional palaeoclimate records. Over the four most recent glacial–interglacial cycles, it is estimated that the British Isles experienced Temperate conditions for only 9% of the time. Over the same period, 60% of Temperate climate episodes were 11 ka or less in length, and 80% were 15 ka or less in length. This analysis confirms the conclusion that 'statistically speaking then, the present interglacial is already on its last legs, tottering along at the advanced age of 10,000 years and can be expected to end within the next 2,000 years' (Imbrie and Imbrie, 1979). Output from the orbital models described in Chapter 9 also indicates that climates as warm as today are relatively rare and suggests that, other things being equal, the global climate should be just beginning a slow deterioration towards glacial conditions. Thus the palaeoclimate and modelling evidence indicates that, with or without enhanced greenhouse warming, the present interglacial will end in the near future.

The description of the sequence of climate change following the end of the Holocene is problematic for two reasons. First, the lack of appropriate high-resolution data limits the detail with which previous interglacial–glacial transitions can be reconstructed (as opposed to the relatively high-resolution record of the last glacial–interglacial transition, i.e. the last 18 ka). Second, the reliability of the last interglacial–glacial transition (about 125 ka to 18 ka BP) as an analogue of the next interglacial–glacial transition is uncertain. The greatest uncertainties concern the effect of enhanced greenhouse warming on the sequence and speed of transition. The post-greenhouse warming transition has effectively no analogue.

Three possible patterns describing the relationship between enhanced greenhouse warming and orbital forcing are identified in Chapter 5. First, the simplest assumption that can be made is of a relatively short (say, 1 ka) period of greenhouse gas-induced warming followed by a return to the 'natural pattern' of glacial–interglacial cycles. The second possibility is that, following a period of warming, the next glaciation will be delayed and will be less severe. The third possibility is that enhanced greenhouse warming will so weaken the positive feedback mechanisms which help to transform the relatively weak orbital forcing into global interglacial–glacial cycles, that the initiation of future glaciation will be prevented. This is the so-called 'irreversible greenhouse effect' and is considered to have a very low, but not zero, probability.

Of the three possibilities, the second, delayed glaciation, is considered most likely. If we further assume that, apart from the effects of greenhouse warming, future climate changes will be similar in their frequency and amplitude to those which have occurred throughout the Pleistocene, then it is possible to generate regional sequences of climate change over the next glacial–interglacial cycle.

In order to simplify the procedure, we represent climate change in a stylized form, as transitions between a set of discrete states. On the basis of the palaeoclimate evidence, four states can be used to describe the range of conditions over the British Isles throughout the Quaternary: Temperate, Boreal, Periglacial or Tundra, and Glacial (see Trewartha, 1968, for a description of their climate characteristics). We have elected to represent the effects of enhanced greenhouse warming as an additional state, Subtropical.

To determine the order in which transitions occur between the climate states, we have used the results from orbital climate models (see Sections 2.3 and 9.3; see also Kukla *et al.*, 1981; Berger, 1988; Tricot *et al.*, 1989; Berger *et al.*, 1991b). Two climate successions have been generated for the next glacial–interglacial cycle, to 125 ka AP. They apply to the northern parts of Britain, which are known to be ice-covered in glacial periods, rather than southern areas, which remain ice-free. Indexes 1 and 2 are shown in Figure 10.1 and represent, respectively, conditions with and without enhanced greenhouse warming. They are only two of the many alternate and equally plausible interpretations of the model and palaeoclimate evidence, but are presented here as a useful basis for the discussion which follows.

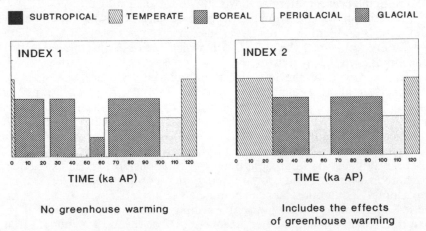

Figure 10.1 Climate indexes showing the succession of major climate states likely to be experienced in the British Isles over the next 125 ka constructed from orbital model output. Index 2 incorporates the potential effects of enhanced greenhouse gas warming.

The output from the orbital models varies from simple climate indexes to estimates of global ice volume. In order to develop successions of future climate from this output it is necessary to establish relationships between past climate conditions in the area of interest and the model reconstructions. If model output consists of global ice volume, for example, we extract the volume estimated for selected events in the past, such as the Holocene thermal optimum, the Last Glacial Maximum and the previous interglacial. When ice volumes of similar magnitude recur in the future, it is assumed that the associated climate conditions in the area of interest will be the same as occurred in the past. Once these basic relationships have been established, the likely range of future conditions can be extrapolated. Inevitably, these procedures entail a large element of subjective judgement.

Index 1 indicates that the present Temperate episode will end within about 2 ka and will be followed by oscillating episodes of colder Boreal and Periglacial climate states leading up to a Glacial episode 52–62 ka AP. This is eventually succeeded by a Temperate interglacial episode (115–125 ka AP).

Index 2 is based on one particular orbital model which includes a crude representation of the enhanced greenhouse effect (Berger *et al.*, 1991b; see also Sections 5.3 and 9.3). This index commences with a period of Subtropical climate representing greenhouse warming, followed by a return to Temperate conditions (1–25 ka AP). In comparison with Index 1, it appears that greenhouse warming may restrict, but not prevent, the expansion of ice sheets in Britain over the next orbital cycle. The Glacial episode of Index 1 (52–62 ka AP) is replaced by a period of Periglacial climate (50–65 ka AP). In this period, ice caps will be present only in the highest, northern mountain areas of Britain, including the Cumbrian Lake District. The climate model on which Index 2 is based only provides simulations for the next 80 ka (Berger *et al.*, 1991b). We, therefore, assume that the pattern of change towards the end of the next glacial– interglacial cycle will be the same as that simulated by other orbital models (i.e. it will be the same as Index 1).

It should be stressed that the Berger model is still in a developmental stage and many uncertainties remain concerning the relationship between enhanced greenhouse warming and orbital forcing mechanisms. Additional uncertainties are introduced by our interpretation of the model results.

If the next glaciation is reduced in extent because of the effects of enhanced greenhouse warming, or is less severe, the pattern of change over the last interglacial–glacial cycle in northern Britain may not provide a reliable guide to future conditions. It could be argued that, in order to investigate conditions during a future reduced glaciation, the magnitude of past changes can simply be scaled down. However, the differential response of the climate system to reduced glaciation is likely to be considerably more complex. This is illustrated by considering the regional response of sea level.

As the deterioration from interglacial to glacial conditions commences, eustatic sea-level fall will dominate. Further into the future, as ice sheets develop in northern Britain, isostatic depression will begin to offset part of the eustatic sea-level fall. If the next glaciation is comparable in magnitude to the late Devensian glaciation, crustal loading by ice will occur well to the south of the Scottish–English border. Eventually, the rate of isostatic depression will exceed the rate of eustatic sea-level fall (Clayton, 1991; Harris, 1991). Thus, at the glacial maximum, regional sea level may be higher than during the preceding period. If, however, the next glaciation is reduced in extent, then the accompanying isostatic depression will be less. If the period of glaciation is also shorter, then a smaller proportion of the total equilibrium depression will occur. Thus *regional* sea-level fall may be greater in a period of reduced glaciation than in the last glaciation, even though the *global* ice volume is less. From the discussion above, it is clear that, in order to determine the sea-level history of a particular site, we need to know its location in relation to the glacial ice-sheet boundaries. If the Cumbrian coastal plain, for example, lies to the south of the ice-sheet margins in the next glaciation, both its glacial and post-glacial history will be very different from that of previous periods.

We conclude that the greatest uncertainties associated with assessments of future long-term climate concern the relationship between anthropogenic greenhouse gas-induced forcing and orbital forcing. At the present time, discussion of this relationship relies on informed judgement and speculation, together with very limited evidence from models. While we can expect considerable improvement in the availability and reliability of model-based evidence, the uncertainties relating to anthropogenic greenhouse gas emissions and to the persistence of the enhanced greenhouse effect are unlikely to be resolved.

These uncertainties are compounded by the fact that, in an attempt to solve one environmental problem, another may be exacerbated. The primary concern relating to the use of CFCs is the depletion of the stratospheric ozone layer. CFCs may, therefore, be replaced by other gases which, although benevolent to stratospheric ozone, contribute to the greenhouse effect. Because of the concern over acid rain, sulphur dioxide emissions from power stations may be reduced. Sulphur dioxide oxidizes to form sulphate aerosols. Once in the stratosphere, these aerosols are very efficient at scattering incoming short-wave radiation back into space, as we observe following major volcanic eruptions. They therefore counteract the effects of greenhouse warming. The reduction of sulphur dioxide emissions (through the introduction of 'clean' technology and/or a reduction in fossil fuel use) may lead to accelerated warming.

10.2 Unexplained variability

The climate models used to construct the regional successions in Section 10.1 are all based on a single forcing mechanism, periodic changes in the orbital parameters. Calibration of model output against the palaeoclimate record indicates that none of these models successfully reproduces *all* the observed variability. The unexplained variance is due, in part, to shortcomings in the parametrization of orbital change mechanisms in the models (Imbrie, 1985). The greater part of the unexplained variance is, however, more a reflection of the fact that climate changes in response to many forcing factors, which operate over many different time scales (see Chapter 1). It has been estimated that orbital-related climatic changes account for only 60% of the variance recorded in the SPECMAP oxygen isotope record over the 19–100 ka part of the frequency spectrum, and for 85% of the variance in the four narrow orbital frequency bands (19 ka, 23 ka, 41 ka and 100 ka) (Imbrie *et al.*, 1984). The remaining variance is accounted for by processes such as solar variability and volcanic activity, together with stochastic mechanisms (see Chapter 3 for a discussion of these mechanisms; see also Mitchell, 1976; Wigley, 1981). Volcanic activity and solar variability are not, at present, predictable, while the stochastic processes cannot, by definition, be predicted.

10.3 Climate transitivity/intransitivity and chaos theory

10.3.1 Intransitivity

It is possible that, given a particular set of boundary conditions, there may be two or more stable sets of climate conditions. In such circumstances, a particular climate state will be stable until an external factor (such as a change in solar radiation) causes a perturbation and the climate moves to a new state. If the forcing is short-lived (such as a volcanic eruption), the original boundary conditions will soon be restored, but the climate may 'relax' into a different stable condition. This type of response is characteristic of an intransitive system, which would be virtually impossible to model (Lorenz, 1968; 1976).

A transitive system, on the other hand, is one where there is only one unique climate state corresponding to a given set of boundary conditions. In this type of system, an external forcing factor will cause a perturbation but, once the external forcing is removed, the climate system will evolve back to the 'normal' climate state.

It is also possible that the climate system is almost intransitive, i.e. it flips between alternate states as minor changes in boundary conditions occur, so that it is almost impossible to identify which is the 'normal' climate state (Lorenz, 1968; 1976; Henderson-Sellers and Robinson, 1986). Such a system would also be extremely difficult to model.

The concept of abrupt climate change has recently gained some popularity (Berger and Labeyrie, 1987). However, considerable confusion surrounds the use of this phrase. Many episodes considered 'abrupt' in the context of the long-term palaeoclimate record, for example, would not be considered 'abrupt' in the context of shorter-term records. Moreover, 'abrupt' changes are less likely to occur in global climate than in regional climate. Climate change cannot be described as 'abrupt' without defining both the temporal and spatial scales concerned.

The Younger Dryas is frequently cited as an example of abrupt climate change (Berger and Labeyrie, 1987; Duplessy *et al.*, 1991). Broecker (1987; Broecker and Denton, 1989) argues that this event represents a flip of the ocean–atmosphere circulation system between two self-stabilizing modes of operation. He suggests that the initiating mechanism is a shutdown of deep-water production in the North Atlantic, which leads to a reorganization of the ocean circulation. A glacial and an interglacial mode of circulation have been identified, but more than two modes may exist (Broecker and Denton, 1989). The modelling evidence in support of ocean circulation mode changes is discussed in Section 2.4.3. The Younger Dryas occurred during a period of warming following the last glaciation. There is no convincing evidence of similar mode changes in periods of generally cooler or cooling climate. This may, however, reflect the more limited availability of appropriate, high-resolution records for such (earlier) periods.

If an almost intransitive response is typical of the climate system, it is possible that increasing atmospheric CO_2 concentrations, for example, will cause the climate to flip to a new state. The existing palaeoclimate records are not, however, sufficiently detailed to determine which type of response is most typical of the Earth's climate. If it turned out to be an intransitive or almost intransitive system, then much of the work on climate modelling would have to be reconsidered.

10.3.2 Chaos theory

Lorenz's ideas relating to intransitive systems have contributed to the development of chaos theory (Gleick, 1988). It is only recently that the implications of this theory for the study of climate dynamics have been considered (Nicolis and Nicolis, 1984; Palmer, 1989; Tsonis and Elsner, 1989; 1990; Lorenz, 1991).

A dynamical system is one whose evolution from some known state can be described by a set of rules or mathematical equations. The evolution of such a system can be represented as a trajectory through n-dimensional 'state' or 'phase' space. According to chaos theory, 'attractors' within state/phase space attract all the trajectories in that space. The simplest type of attractor is a point. In such a system, for example a pendulum, a point of no motion is eventually reached. Thus, the long-term evolution of the system is not sensitive to its initial condition. The system is non-chaotic and predictable. However, other systems are characterized by more complex 'strange attractors'. In this case, the evolution of the system from two slightly different initial conditions will be completely different. The trajectory never crosses itself and never goes back on itself. This is the so-called *Lorenz attractor* and has a characteristic butterfly-wing shape (Figure 10.2). Thus, although the system can be described by simple deterministic rules, it can generate randomness or chaos. If the initial conditions are known exactly, the future evolution of the system can be predicted by tracing its trajectory through space.

An attractor has a dimension which indicates the minimum number of variables responsible for its evolution. If the number of variables is known and an exact mathematical formulation of the system can be developed, it is theoretically possible to generate the system trajectories in space and to

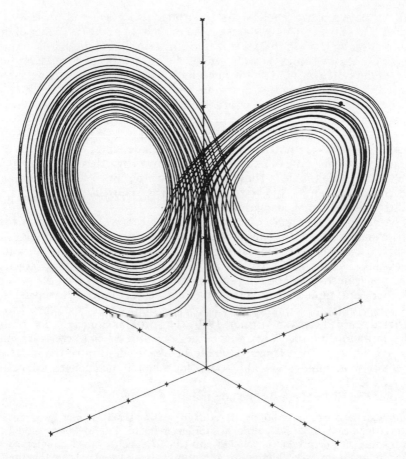

Figure 10.2 The Lorenz attractor. Figure provided by J.P. Crutchfield.

identify the attractor(s). Because exact mathematical formulations of the atmosphere or climate system are not available, it is more appropriate to use observed weather/climate variables in non-linear differential equations to investigate whether or not the system is likely to display chaotic behaviour (Palmer, 1989; Tsonis and Elsner, 1989). A number of studies have used this approach to identify attractors in the weather and climate system (Nicolis and Nicolis, 1984; Fraedrich, 1986; Essex *et al.*, 1987; Tsonis and Elsner, 1988). Most of these have considered short-term variability only, but the earliest study identified a climate attractor with a dimension of 3.1 in the oxygen isotope record of the last 1 Ma. This compares with dimensions of 6–7 estimated from short-term records (50 a to 11 hours).

Tsonis and Elsner (1990) have investigated the existence of multiple attractors and long-term climate dynamics using a very simple model which describes forced, damped, non-linear oscillations of the climate system in response to internal forcing. The system contains two periodic attractors and oscillates between two stable states (identified as ice age and present-day modes). These two modes or states can be thought of as the two wings of the Lorenz attractor (Figure 10.2). Depending on its initial condition, the model climate eventually settles down to one of the attractors. The trajectory, or orbit, of the system around this attractor will at times come very close to the trajectory of the second attractor. Tsonis and Elsner argue that the existence of noise (related to short-term internal and/or external fluctuations) may be strong enough to cause the system to jump into the trajectory of the second attractor. This behaviour is a frequently observed property of chaotic dynamical systems (Lorenz, 1991). That is, extended intervals in which the system exhibits one type of behaviour are followed by extended intervals in which another dominates. These intervals may be much longer than any time constants identified in the governing laws. Thus, it is impossible to predict when the mode changes will occur.

If we accept the existence of chaos and the presence of strange attractors in the climate system, does this make prediction impossible? Certainly, the sensitive dependence on initial conditions poses problems for weather forecasting (Palmer, 1989). Unless the initial conditions are known exactly, the wrong trajectory will be followed: the predicted evolution of the system will, therefore, rapidly diverge from the 'correct' evolution. However, the presence of attractors implies that future variability will be confined to a restricted space (Tsonis and Elsner, 1989). The problem then is to determine which attractors are involved and whether or not their characteristics may change (Palmer, 1989). Thus, we would be able to determine the range of variability, but would not be able to predict when mode changes will occur (Lorenz, 1991).

Tsonis and Elsner (1989) conclude that

> although the existence or not of attractors in weather is still an open question, it is significant that the empirical studies indicate that low-dimensional attractors may be present in weather and climate. The fact that the inferred dimensions seem to be different for different timescales may indicate that the attractors (and thus predictability) are different and a function of the timescale.

If this view is correct, chaos theory may help us to investigate the predictability and the interaction of climate changes on different time scales.

10.4 Concluding remarks

Despite all the uncertainties, we consider that a firm basis for the assessment of future long-term climate change can be established. For climate change over the next 1 Ma, this basis is provided by the palaeoclimate record of the last 1 Ma and by models. The use of regional palaeoclimate records allows us to say something about future regional, as well as global, change. While recognizing the current limitations of our knowledge, we consider that climatologists can provide useful information for input to sensitivity studies and long-term impact assessments (see, for example, Goodess *et al.*, 1992).

Continuing advances in our knowledge, based on modelling, data collection and analysis, are expected to be made in the coming years. A number of major global research programs are under way, or are planned, and should provide valuable information. As we find out more about the climate system, however, it sometimes seems that more questions are raised than answered. Our own ability to modify the climate, either deliberately or inadvertently, is likely to become an increasingly important consideration. Over the very long future time scales considered in this book, we can only begin to guess at the forms these modifications might take. Thus, we expect the future of climate research to be exciting and challenging, with some potential surprises.

References

Adem, J. (1988) Possible causes and numerical simulation of the Northern Hemispheric climate during the last deglaciation. *Atmosfera*, *1*, 17–38.

Adem, J. (1989) On the effect of the orbital variation on the climates from 4 thousand years ago to present. *Annales Geophysicae*, *7*(6), 599–606.

Anderson, J.B. and Thomas, M.A. (1991) Marine ice-sheet decoupling as a mechanism for rapid episodic sea-level change: the record of such events and their influence on sedimentation. *Sedimentary Geology*, *70*(2–4), 87–104.

Atkinson, T.C., Briffa, K.R. and Coope, G.R. (1987) Seasonal temperatures in Britain during the past 22,000 years, reconstructed using beetle remains. *Nature*, *325*(6105), 587–592.

Baillie, M.G.L. and Munro, M.A.R. (1988) Irish tree rings, Santorini and volcanic dust veils. *Nature*, *332*(6162), 344–346.

Bard, E., Arnold, M., Maurice, P., Duprat, J., Moyes, J. and Duplessy, J.C. (1987) Retreat velocity of the North Atlantic polar front during the last deglaciation determined by ^{14}C accelerator mass spectrometry. *Nature*, *328*(6133), 791–793.

Bard, E., Fairbanks, R., Arnold, M., Maurice, P., Duprat, J., Moyes, J. and Duplessy, J.C. (1989) Sea-level estimates during the last deglaciation based on ^{18}O and accelerator mass spectrometry ^{14}C ages measured in Globigerina bulloides. *Quaternary Research*, *31*(3), 381–391.

Bard, E., Hamelin, B. and Fairbanks, R.G. (1990) U–Th ages obtained by mass spectrometry in corals from Barbados: sea level during the past 130,000 years. *Nature*, *346*(6283), 456–458.

Barnett, T.P. (1983) Recent changes in sea level and their possible causes. *Climatic Change*, *5*(1), 15–38.

Barnett, T.P. (1984) The estimation of 'global' sea level change: a problem of uniqueness. *Journal of Geophysical Research*, *89*(C5), 7980–7988.

Barnett, T.P. (1985) On long-term climatic changes in observed physical properties of the oceans. In M.C. MacCracken and F.M. Luther (eds), *Detecting the Climatic Effects of Increasing Carbon Dioxide*, DOE/ER-0235, Carbon Dioxide Research Division, Office of Energy Research, US Department of Energy, Washington, DC, pp. 91–107.

Barnett, T.P. (1988) Global sea level change. In NCPO, *Climate Variations over the Past Century and the Greenhouse Effect: A Report Based on the First Climate Trends Workshop*, 7–9 September 1988, Washington, DC. National Climate Program Office/NOAA, Rockville, Maryland.

Barnett, T.P. (1989) A solar–ocean relation—fact or fiction? *Geophysical Research Letters*, *16*(8), 803–806.

Barnola, J.M., Raynaud, D., Korotkevich, Y.S. and Lorius, L. (1987) Vostock ice core provides 160,000-year record of atmospheric CO_2. *Nature*, *329*(6138), 408–414.

Barron, E.J. (1988) Studies of Cretaceous climate. In A. Berger, R. Dickinson, and J. Kidson (eds), *Contributions of Geophysics to Climate Change Studies*, American Geophysical Union, Washington, DC.

Barry, R.G. (1985) The cryosphere and climate change. In M.C. MacCracken and F.M. Luther (eds) *Detecting the Climatic Effects of Increasing Carbon Dioxide*,

DOE/ER-0235, Carbon Dioxide Research Division, Office of Energy Research, US Department of Energy, Washington, DC, pp. 109–148.

Bates, T.S., Charlson, R.J. and Gammon, R.H. (1987) Evidence for the climatic role of marine biogenic sulfur. *Nature*, *329*(6137), 319–321.

Beer, J., Siegenthaler, U., Bonani, G., Finkel, R.C., Oeschger, H., Suter, M. and Wölfli, W. (1988) Information on past solar activity and geomagnetism from ^{10}Be in the Camp Century ice core. *Nature*, *331*(6158), 675–679.

Beget, J.E. and Hawkins, D.B. (1989) Influence of orbital parameters on Pleistocene loess deposition in central Alaska. *Nature*, *337*(6203), 151–153.

Berger, A. (1977) Long-term variation of the Earth's orbital elements. *Celestial Mechanics*, *15*, 53–74.

Berger, A. (1978a) Long-term variations of caloric insolation resulting from the Earth's orbital elements. *Quaternary Research*, *9*(2), 139–167.

Berger, A. (1978b) Long-term variations of daily insolation and Quaternary climatic changes. *Journal of the Atmospheric Sciences*, *35*(12), 2362–2367.

Berger, A. (1984) A critical review of modeling the astronomical theory of paleoclimates and the future of our climate. In *Proceedings of the International Conference on Sun and Climate*, CNES-CNRS-DGRST, Toulouse, October, 1984, pp. 325–355.

Berger, A. (1988) Milankovitch theory and climate. *Reviews of Geophysics*, *26*(4), 624–657.

Berger, A. (1989a) Pleistocene climatic variability at astronomical frequencies. *Quaternary International*, *2*, 1–14.

Berger, A. (1989b) The spectral characteristics of pre Quaternary climatic records, an example of the relationship between the astronomical theory and geo-sciences. In A. Berger, S. Schneider and J.C. Duplessy (eds), *Climate and Geo-Sciences*. Kluwer Academic Publishers, Dordrecht, pp. 47–76.

Berger, A. and Pestiaux, P. (1984) Accuracy and stability of the Quaternary terrestrial insolation. In A. Berger, J. Imbrie, J. Hays, G. Kukla and B. Saltzman (eds), *Milankovitch and Climate*, D. Reidel, Dordrecht, pp. 83–111.

Berger, A. and Tricot, C. (1986) Global climatic changes and astronomical theory of paleoclimates. In A. Cazenave (ed.), *Earth Rotation: Solved and Unsolved Problems*, D. Reidel, Dordrecht. pp. 111–129.

Berger, A., Guiot, J., Kukla, G. and Pestiaux, P. (1981) Long-term variation of monthly insolation as related to climatic changes. *Sonderdruck aus der Geologischen Rundschau*, *70*(2), 748–758.

Berger, A., Imbrie, J., Hays, J., Kukla, G. and Saltzman, B. (eds) (1984) *Milankovitch and Climate. Parts 1–2. Understanding the Response to Astronomical Forcing*, 2 vols, D. Reidel, Dordrecht.

Berger, A., Loutre, M.F. and Dehant, V. (1989a) Influence of the changing lunar orbit on the astronomical frequencies of pre-Quaternary insolation patterns. *Paleoceanography*, *4*(5), 555–564.

Berger, A., Loutre, M.F. and Dehant, V. (1989b) Pre-Quaternary Milankovitch frequencies. *Nature*, *342*(6246), 133.

Berger, A., Gallée, H., Fichefet, T., Marsiat, I. and Tricot, C. (1990) Testing the astronomical theory with a coupled climate-ice sheet model. *Palaeogeography, Palaeoclimatology, Palaeoecology (Global and Planetary Change Section)*, *89*, 125–141.

Berger, A., Fichefet, T., Gallée, H., Marsiat, I., Tricot, C. and Van Ypersele, J.P. (1991a) Physical interactions within a coupled climate model over the last glacial interglacial cycle. *Transactions of the Royal Society of Edinburgh — Earth Sciences*, *81*(4), 357–370.

References

Berger, A., Gallée, H. and Melice, J.L. (1991b) The Earth's future climate at the astronomical time scale. In C.M. Goodess and J.P. Palutikof, (eds), *Proceedings of the International Workshop on Future Climate Change and Radioactive Waste Disposal*, Nirex Safety Series, NSS/R257, pp. 148–165.

Berger, W.H. (1990) The Younger Dryas cold spell — a quest for causes. *Palaeogeography, Palaeoclimatology, Palaeoecology (Global and Planetary Change Section)*, *89*(3), 219–237.

Berger, W.H. and Labeyrie, L.D. (eds) (1987) *Abrupt Climatic Change: Evidence and Implications*. NATO Advanced Study Institute Series, Series C, Mathematical and Physical Sciences, Vol. 216. D. Reidel, Dordrecht.

Berner, W., Oeschger, H. and Stauffer, B. (1980) Information on the CO_2 cycle from ice core studies. *Radiocarbon*, *22*(2), 227–234.

Birchfield, G.E., Weertman, J. and Lunde, A.T. (1981) A palaeoclimate model of Northern Hemisphere ice sheets. *Quaternary Research*, *15*(2), 126–142.

Birks, H.J.B. and Peglar, S.M. (1979) Interglacial pollen from Sel Ayre, Shetland. *New Phytologist*, *83*, 559–575.

Bishop, W.W. and Coope, G.R. (1977) Stratigraphical and faunal evidence for lateglacial and early Flandrian environments in south-west Scotland. In J.M. Gray and J.J. Lowe (eds), *Studies in the Scottish Lateglacial Environment*, Pergamon Press, Oxford, pp. 61–88.

Bjerknes, J. (1968) Atmosphere–ocean interaction during the 'Little Ice Age' (seventeenth to nineteenth centuries AD). *Meteorological Monographs*, *8*, 37–62.

Bloemendal, J. and de Menocal, P. (1989) Evidence for a change in the periodicity of tropical climate cycles at 2.4 Myr from whole-core magnetic susceptibility measurements. *Nature*, *342*(6251), 897–899.

Bloom, A.L. (1970) Holocene submergence in Micronesia as the standard for eustatic sea-level changes. In *Symposium on the Evolution of Shorelines and Continental Shelves during the Quaternary*. *Quaternaria*, 12, 145–154.

Bloom, A.L., Broecker, W.S., Chappell, J.M.A., Matthews, R.K. and Mesolella, K.J. (1974) Quaternary sea level fluctuations on a tectonic coast: new $^{230}Th/^{234}U$ dates from the Huon Peninsula, New Guinea. *Quaternary Research*, *4*(2), 85–205.

Bolin, B., Doos, B.R., Jäger, J. and Warrick, R.A. (eds) (1986) *The Greenhouse Effect, Climatic Change, and Ecosystems*. Scientific Committee on Problems of the Environment (SCOPE) of the International Council of Scientific Unions, SCOPE 29. Wiley, New York.

Boulton, G.S. and Clark, C.D. (1990) A highly mobile Laurentide ice sheet revealed by satellite images of glacial lineations. *Nature*, *346*(6287), 813–817.

Boulton, G.S., Jones, A.S., Clayton, K.M. and Kenning, M.J. (1977) A British ice sheet model and patterns of glacial erosion and deposition in Britain. In F.W. Shotton (ed.), *British Quaternary Studies*, Oxford University Press, Oxford, pp. 231–246.

Boulton, G.S., Smith, G.D., Jones, A.S. and Newsome, J. (1985) Glacial geology and glaciology of the last mid-latitude ice sheets. *Journal of the Geological Society*, *142*(3), 447–474.

Bowen, D.Q. (1978) *Quaternary Geology: a Stratigraphic Framework for Multidisciplinary Work*. Pergamon Press, Oxford.

Bowen, D.Q. and Sykes, G.A. (1988) Correlation of marine events and glaciations on the northeast Atlantic margin. *Philosophical Transactions of the Royal Society of London B*, *318*, 619–635.

Bowen, D.Q., Sykes, G.A., Reeves, A., Miller, G.H., Andrews, J.T., Brew, J.S. and Hare, P.E. (1985) Amino acid geochronology of raised beaches in south west Britain. *Quaternary Science Reviews*, *4*, 279–318.

Bowen, D.Q., Rose, J., McCabe, A.M. and Sutherland, D.G. (1986) Correlation of Quaternary glaciations in England, Ireland, Scotland and Wales. *Quaternary Science Reviews*, *5*, 299–340.

Bowen, D.Q., Hughes, S., Sykes, G.A. and Miller, G.H. (1989) Land–sea correlations in the Pleistocene based on isoleucine epimerization in non-marine molluscs. *Nature*, *340*(6228), 49–51.

Boyle, E.A. (1988) Vertical oceanic nutrient fractionation and glacial/interglacial CO_2 cycles. *Nature*, *331*(6151), 55–56.

Boyle, E.A. and Keigwin, L. (1987) North Atlantic thermohaline circulation during the past 20,000 years linked to high-latitude surface temperature. *Nature*, *330*(6143), 35–40.

Boyle, E.A. and Rosener, P. (1990) Further evidence for a link between late Pleistocene North Atlantic surface temperatures and North Atlantic deep-water production. *Palaeogeography, Palaeoclimatology, Palaeoecology (Global and Planetary Change Section)*, *89*(1–?), 113–124.

Bradley, R.S. (1985) *Quaternary Paleoclimatology, Methods of Paleoclimatic Reconstruction*, Allen & Unwin, London.

Bradley, R.S. (1988) The explosive volcanic eruption signal in Northern Hemisphere continental temperature records. *Climatic Change*, *12*(3), 221–244.

Bretherton, F.P., Bryan, K. and Woods, J.D. (1990) Time-dependent greenhouse-gas-induced climate change. In J.T. Houghton, G.J. Jenkins and J.J. Ephraums (eds), *Climate Change, the IPCC Scientific Assessment*, WMO/UNEP, Cambridge University Press, Cambridge, pp. 173–194.

Briffa, K.R., Bartholin, T.S., Eckstein, D., Jones, P.D., Karlén, W., Schweingruber, F.H. and Zetterberg, P. (1990) A 1,400-year tree-ring record of summer temperatures in Fennoscandia. *Nature*, *346*(6283), 434–439.

Broecker, W.S. (1981) Glacial to interglacial changes in ocean and atmospheric chemistry. In A. Berger (ed.), *Climatic Variations and Variability: Facts and Theories*, D. Reidel, Dordrecht, pp. 109–120.

Broecker, W.S. (1982) Glacial to interglacial changes in ocean chemistry. *Progress in Oceanography*, *11*, 151–197.

Broecker, W.S. (1987) Unpleasant surprises in the greenhouse? *Nature*, *328*(6126), 123–126.

Broecker, W.S. and Denton, G.H. (1989) The role of ocean–atmosphere reorganizations in glacial cycles. *Geochimica et Cosmochimica Acta*, *53*(10), 2465–2502.

Broecker, W.S. and Denton, G.H. (1990) What drives glacial cycles? *Scientific American*, *262*(1), 43–50.

Broecker, W.S., Peteet, D.M. and Rind, D. (1985) Does the ocean–atmosphere system have more than one stable mode of operation? *Nature*, *315*(6014), 21–25.

Broecker, W.S., Andree, M., Bonani, G., Wolfli, W., Oeschger, H. and Klas, M. (1988) Can the Greenland climatic jumps be identified in records from ocean and land? *Quaternary Research*, *30*(1), 1–6.

Bryson, R.A. (1989) Late Quaternary volcanic modulation of Milankovitch climate forcing. *Theoretical and Applied Climatology*, *39*(3), 115–125.

Budd, W.F., McInnes, B.J., Jenssen, D. and Smith, I.N. (1987) Modelling the response of the West Antarctic ice sheet to a climatic warming. In C.J. Van der Veen and J. Oerlemans (eds), *Dynamics of the West Antarctic Ice Sheet*, Workshop on the Dynamics of the West Antarctic Ice Sheet, University of Utrecht, Netherlands, 6–8 May, D. Reidel, Dordrecht, pp. 321–358.

Budyko, M.I., Ronov, A.B. and Yanshin, A.L. (1987) *History of the Earth's Atmosphere* (English translation), Springer-Verlag, Berlin.

References

Calder, N. (1974) Arithmetic of ice ages. *Nature, 252*(5480), 216–218.

Castagnoli, G.C., Bonino, G., Caprioglio, F., Serio, M., Provenzale, A. and Bhandari, N. (1990) The $CaCO_3$ profile in a recent Ionian Sea core and the tree ring radiocarbon record over the last 2 millennia. *Geophysical Research Letters, 17*(10), 1545–1548.

Cess, R.D. (1989) Greenhouse effect: gauging water-vapour feedback. *Nature, 342*(6251), 736–737.

Cess, R.D. (1991) Greenhouse warming — positive about water feedback. *Nature, 349*(6309), 462.

Cess, R.D. and Potter, G.L. (1988) A methodology for understanding and intercomparing atmospheric climate feedback processes in general circulation models. *Journal of Geophysical Research, 93*(D7), 8305–8314.

Cess, R.D., Potter, G.L., Blanchet, J.P., Boer, G.J., Ghan, S.J., Kiehl, J.T., Le Treut, H., Li, Z.X., Liang, X.Z., Mitchell, J.F.B., Morcrette, J.J., Randall, D.A., Riches, M.R., Roeckner, E., Schlese, U., Slingo, A., Taylor, K.E., Washington, W.M., Wetherald, R.T. and Yagai, I. (1989) Interpretation of cloud–climate feedback as produced by 14 atmospheric general circulation models. *Science, 245,* 513–515.

Cess, R.D., Potter, G.L., Blanchet, J.P., Boer, G.J., Del Genio, A.D., Déqué, M., Dymnikov, V., Galin, V., Gates, W.L., Ghan, S.J., Kiehl, J.T., Lacis, A.A., Le Treut, H., Li, Z.X., Liang, X.Z., McAvaney, B.J., Meleshko, V.P., Mitchell, J.F.B., Morcrette, J.J., Randall, D.A., Rikus, L., Roeckner, E., Royer, J.F., Schlese, U., Sheinin, D.A., Slingo, A., Sokolov, A.P., Taylor, K.E., Washington, W.M., Wetherald, R.T., Yagai, I. and Zhang, M.H. (1990) Intercomparison and interpretation of climate feedback processes in 19 atmospheric general circulation models. *Journal of Geophysical Research, 95*(D10), 16601–16615.

Cess, R.D., Potter, G.L., Zhang, M.H., Blanchet, J.P., Chalita, S., Colman, R., Dazlich, D.A., Del Genio, A.D., Dymnikov, V., Galin, V., Jerrett, D., Keup, E., Lacis, A.A., Le Treut, H., Liang, X.Z., Mahfouf, J.F., McAvaney, B.J., Meleshko, V.P., Mitchell, J.F.B., Morcrette, J.J., Norris, P.M., Randall, D.A., Rikus, L., Roeckner, E., Royer, J.F., Schlese, U., Sheinin, D.A., Slingo, J.M., Sokolov, A.P., Taylor, K.E., Washington, W.M., Wetherald, R.T. and Yagai, I. (1991) Interpretation of snow–climate feedback as produced by 17 general circulation models. *Science, 253*(5022), 888–892.

Chapellaz, J., Barnola, J.M., Raynaud, D., Korotkevich, Y.S. and Lorius, C. (1990) Ice-core record of atmospheric methane over the past 160,000 years. *Nature, 345*(6270), 127–131.

Chappell, J. and Shackleton, N.J. (1986) Oxygen isotopes and sea level. *Nature, 324*(6093), 137–140.

Charles, C.D. and Fairbanks, R.G. (1992) Evidence from Southern Ocean sediments for the effect of North Atlantic deep-water flux on climate. *Nature, 355*(6359), 416–419.

Charlock, T.P., Cattany-Carnes, K.M. and Rose, F. (1988) Fluctuation statistics of outgoing longwave radiation in a general circulation model and in satellite data. *Monthly Weather Review, 116*(8), 1540–1554.

Charlson, R.J., Lovelock, J.E., Andreae, M.O. and Warren, S.G. (1987) Oceanic phytoplankton, atmospheric sulphur, cloud albedo and climate. *Nature, 326*(6114), 655–661.

Chen, J.H., Curran, H.A., White, B. and Wasserburg, G.J. (1991) Precise chronology of the last interglacial period — U^{234}–Th^{230} data from fossil coral reefs in the Bahamas. *Geological Society of America Bulletin, 103*(1), 82–97.

Chen, R.S. and Robinson, P.J. (1991) Generating scenarios of local surface temperature using time series methods. *Journal of Climate, 4*(7), 723–732.

Clayton, K.M. (1991) Long-term sea-level change. In C.M. Goodess and J.P. Palutikof (eds) *Proceedings of the International Workshop on Future Climate Change and Radioactive Waste Disposal,* Nirex Safety Series, NSS/R257, pp.191–202.

CLIMAP Project Members (1976) The surface of the ice-age Earth. *Science, 191,* 1131–1137.

CLIMAP Project Members (1981) Seasonal reconstructions of the Earth's surface at the last glacial maximum. *Geological Society of America Map Chart Series, MC-36,* pp. 1–18.

CLIMAP Project Members (1984) The last interglacial ocean. *Quaternary Research, 21*(2), 123–224.

COHMAP Members (1988) Climatic changes of the last 18,000 years: observations and model simulations. *Science, 241,* 1043–1052.

Conolly, A.P. and Dahl, F. (1970) Maximum summer temperature in relation to the modern and Quaternary distributions of certain arctic-montane species in the British Isles. In D. Walker and R.G. West (eds), *Studies in the Vegetational History of the British Isles,* Cambridge University Press, Cambridge, pp. 159–224.

Coope, G.R. (1975) Climatic fluctuations in northwest Europe since the last interglacial, indicated by fossil assemblages of Coleoptera. In A.E. Wright and F. Moseley (eds), *Ice Ages: Ancient and Modern,* Seel House Press, Liverpool, pp. 153–168.

Coope, G.R., Morgan, A. and Osborne, P.J. (1971) Fossil Coleoptera as indicators of climatic fluctuations during the last glaciation in Britain. *Palaeogeography, Palaeoclimatology, Palaeoecology, 10,* 87–101.

Covey, C., Taylor, K.E. and Dickinson, R.E. (1991) Upper limit for sea ice albedo feedback contribution to global warming. *Journal of Geophysical Research, 96*(D5), 9169–9174.

Croll, J. (1875) *Theory of Secular Changes of the Earth Climate, Climate and Time in their Geological Relations,* Appleton and Co., New York.

Crowley, T.J. (1983) The geologic record of climatic change. *Reviews of Geophysics and Space Physics, 21*(4), 828–877.

Cubasch, U. and Cess, R.D. (1990) Processes and modelling. In J.T. Houghton, G.J. Jenkins and J.J. Ephraums (eds), *Climate Change, the IPCC Scientific Assessment,* WMO/UNEP, Cambridge University Press, Cambridge, pp. 69–92.

Cwynar, L.C. and Watts, W.A. (1989) Accelerator-mass spectrometer ages for late-glacial events at Ballybetagh, Ireland. *Quaternary Research, 31*(3), 377–380.

Dansgaard, W. and Tauber, H. (1969) Glacier oxygen-18 and Pleistocene ocean temperatures. *Science, 166,* 499–502.

Dansgaard, W., White, J.W.C. and Johnsen, S.J. (1989) The abrupt termination of the Younger Dryas climate event. *Nature, 339*(6225), 532–534.

de Beaulieu, J.L., Guiot, J. and Reille, M. (1991) Long European pollen records and quantitative reconstructions of the last climatic cycle. In C.M. Goodess and J.P. Palutikof (eds) *Proceedings of the International Workshop on Future Climate Change and Radioactive Waste Disposal,* Nirex Safety Series, NSS/R257, pp. 116–136.

Deblonde, G. and Peltier, W.R. (1991a) Simulations of continental ice sheet growth over the last glacial–interglacial cycle: experiments with a one-level seasonal energy balance model including realistic geography. *Journal of Geophysical Research, 96*(D5), 9189–9215.

Deblonde, G. and Peltier, W.R. (1991b) A one-dimensional model of continental ice volume fluctuations through the Pleistocene: implications for the origin of the mid-Pleistocene climate transition. *Journal of Climate, 4*(3), 318–344.

References

De Deckker, P., Correge, T. and Head, J. (1991) Late Pleistocene record of cyclic eolian activity from tropical Australia suggesting the Younger Dryas is not an unusual climatic event. *Geology, 19,* 602–605.

Deepak, A. and Gerber, H.E. (eds) (1983) *Report of the Experts Meeting on Aerosols and their Climatic Effects,* WCP-55, World Meteorological Organization, Geneva.

Delcourt, P.A. and Delcourt, H.R. (1984) Late Quaternary paleoclimates and biotic responses in eastern North America and the western North Atlantic Ocean. *Palaeogeography, Palaeoclimatology, Palaeoecology, 48*(2–4), 263–284.

Del Genio, A.D., Lacis, A.A. and Ruedy, R.A. (1991) Simulations of the effect of a warmer climate on atmospheric humidity. *Nature, 351*(6325), 382–385.

Delmas R.J., Ascencio, J.-M. and Legrand, M. (1980) Polar ice evidence that atmospheric CO_2 20,000 BP was 50% of present. *Nature,* 284(5752), 155–157.

Denton, G.H. and Hughes, T.J. (1983) Milankovitch theory of ice ages: hypothesis of ice-sheet linkage between regional insolation and global climate. *Quarterly Research, 20*(2), 125–145.

Denton, G.H. and Karlén, W. (1973) Holocene climatic variations — their pattern and possible cause. *Quaternary Research, 3*(2), 155–205.

Denton, G.H., Hughes, T.J. and Karlén, W. (1986) Global ice-sheet system interlocked by sea level. *Quaternary Research, 26*(1), 3–26.

Dickinson, R.E., Meehl, G.A. and Washington, W.M. (1987) Ice–albedo feedback in a CO_2-doubling simulation. *Climatic Change, 10*(3), 241–248.

Dickinson, R.E. and Henderson-Sellers, A. (1988) Modelling tropical deforestation: a study of GCM land-surface parametrizations. *Quaternary Journal of the Royal Meteorological Society, 114*(480), 439–462.

Dowsett, H.J. and Poore, R.Z. (1990) A new planktic foraminifer transfer function for estimating Pliocene–Holocene paleoceanographic conditions in the North Atlantic. *Marine Micropaleontology, 16*(1–2), 1–23.

Duplessy, J.C. and Juillet-Leclerc, A. (1991) Global ocean circulation during the last ice age: enhanced intermediate water formation and reduced deep water circulation. In C.M. Goodess and J.P. Palutikof, (eds), *Proceedings of the International Workshop on Future Climate Change and Radioactive Waste Disposal,* Nirex Safety Series, NSS/R257, pp. 106–115.

Duplessy, J.C. and Shackleton, N.J. (1985) Response of global deep-water circulation to Earth's climatic change 135,000–107,000 years ago. *Nature, 316*(6028), 500–506.

Duplessy, J.C., Delibrias, G., Turon, J.L., Pujol, C. and Duprat, J. (1981) Deglacial warming of the northeastern Atlantic Ocean: correlation with the paleoclimatic evolution of the European continent. *Palaeogeography, Palaeoclimatology, Palaeoecology, 35,* 121–144.

Duplessy, J.C., Arnold, M., Maurice, P., Bard, E., Duprat, J. and Moyes, J. (1986) Direct dating of the oxygen-isotope record of the last deglaciation by ^{14}C accelerator mass spectrometry. *Nature, 320*(6060), 350–352.

Duplessy, J.C., Labeyrie, L., Kallel, N. and Juillet- Leclerc, A. (1989) Intermediate and deep water characteristics during the last glacial maximum. In A. Berger, S. Schneider and J.C. Duplessy (eds), *Climate and Geo-Sciences,* Kluwer Academic Publishers, Dordrecht, pp. 105–120.

Duplessy, J.C., Bard, E., Arnold, M., Shackleton, N.J., Duprat, J. and Labeyrie, L. (1991) How fast did the ocean–atmosphere system run during the last deglaciation? *Earth and Planetary Science Letters, 103*(1–4), 27–40.

Eddy, J.A. (1976) The Maunder Minimum. *Science, 192,* 1189–1202.

Eddy, J.A. (1977) Climate and the changing sun. *Climatic Change, 1*(2), 173–190.

Eddy, J.A. (1982) The solar constant and surface temperature. In R.A. Peck and

J.R. Hummel. (eds), *Interpretation of Climate and Photochemical Models, Ozone and Temperature Measurements,* American Institute of Physics, New York.

Ekman, M. (1988) The world's longest continued series of sea level observations. *Pure and Applied Geophysics, 127*(1), 73–77.

Emery, K.O. (1980) Relative sea levels from tide-gauge records. *Proceedings of the National Academy of Science, 77,* 6968–6972.

Emery, K.O. and Aubrey, D.G. (1985) Glacial rebound and relative sealevels in Europe from tide-gauge records. *Tectonophysics, 120,* 239–255.

Emiliani, C. (1955) Pleistocene temperatures. *Journal of Geology, 63,* 538–578.

Emiliani, C. (1966) Isotopic palcotemperatures. *Science, 154*(3750), 851–857.

Entekhabi, D. and Eagleson, P.S. (1989) Land surface hydrology parameterization for atmospheric general circulation models including subgrid scale spatial variability. *Journal of Climate, 2*(8), 816–831.

Essex, C., Lookman, T. and Neremberg, M.A.H. (1987) The climate attractor over short time scales. *Nature, 326*(6108), 64–66.

Fairbanks, R.G. (1989) A 17,000-year glacio-eustatic sea level record: influence of glacial melting rates on the Younger Dryas event and deep-ocean circulation. *Nature, 342*(6250), 637–642.

Fairbridge, R.W. (1961) Eustatic changes in sca level. *Physics and Chemistry of the Earth, 4,* 99–185.

Fairbridge, R.W. and Krebs, O. (1962) Sea level and the Southern Oscillation. *Geophysical Journal of the Royal Astronomical Society, 6,* 532–545.

Fischer, A.G. (1986) Climatic rhythms recorded in strata. *Annual Review of Earth and Planetary Science,* 14, 351–376.

Fisher, D.A., Hales, C.H., Wang, W.C., Ko, M.K.W. and Sze, N.D. (1990) Model calculations of the relative effects of CFCs and their replacements on global warming. *Nature, 344*(6266), 513–516.

Flashchka, I.M., Stockton, C.W. and Boggess, W.R. (1987) Climatic variation and surface water resources in the Great Basin region. *Water Resources Bulletin, 23*(1), 47–57.

Flinn, D. (1981) A note on the glacial and late glacial history of Caithness. *Geological Journal, 16,* 175–179.

Flohn, H. (1983) Actual palaeoclimatic problems from a climatologists viewpoint. In A. Ghazi (ed.), *Palaeoclimatic Research and Models,* D. Reidel, Dordrecht, pp. 17–33.

Flohn, H. and Fantechi, R. (1984) *The Climate of Europe: Past, Present and Future.* D. Reidel, Dordrecht.

Foley, J.A., Taylor, K.E. and Ghan, S.J. (1991) Planktonic dimethylsulfide and cloud albedo — an estimate of the feedback response. *Climatic Change, 18*(1), 1–15.

Folland, C.K., Parker, D.E. and Kates, F.E. (1984) Worldwide marine temperature fluctuations 1856–1981. *Nature, 310*(5979), 670– 673.

Folland, C.K., Karl, T. and Vinnikov, K.Ya. (1990) Observed climate variations and change. In J.T. Houghton, G.J. Jenkins and J.J. Ephraums (eds), *Climate Change, the IPCC Scientific Assessment,* WMO/UNEP, Cambridge University Press, Cambridge, pp. 195–238.

Foukal, P.V. (1990) The variable sun. *Scientific American, 262*(2), 34–41.

Fraedrich, K. (1986) Estimating the dimensions of weather and climate attractors. *Journal of Atmospheric Sciences, 44,* 419–432.

Frakes, L.A. (1979) *Climates throughout Geological Time.* Elsevier, Amsterdam.

Fullerton, D.S. and Richmond, G.M. (1986) Comparison of the marine oxygen isotope record, the eustatic sea level record, and the chronology of glaciation in the United States of America. *Quaternary Science Reviews, 5,* 197–200.

References

Funnell, B.M. (1991) Geological evidence from deep sea cores. In C.M. Goodess and J.P. Palutikof (eds), *Proceedings of the International Workshop on Future Climate Change and Radioactive Waste Disposal*, Nirex Safety Series, NSS/R257, pp. 95–103.

Gallée, H., van Ypersele, J.P., Fichefet, T., Tricot, C. and Berger, A. (1991) Simulation of the last glacial cycle by a coupled, sectorially averaged climate-ice sheet model. 1. The climate model. *Journal of Geophysical Research, 96*(D7), 13139–13161.

Gascoyne, M., Currant, A.P. and Lord, T.C. (1981) Ipswichian fauna of Victoria Cave and the marine palaeoclimatic record. *Nature, 294*(5842), 652–654.

Gasse, F., Ledee, V., Massault, M. and Fontes, J-C. (1989) Water-level fluctuations of Lake Tanganyika in phase with oceanic changes during the last glaciation and deglaciation. *Nature, 342*(6245), 57–59.

Gates, W.L. (1985) The use of general circulation models in the analysis of the ecosystem impacts of climatic change. *Climatic Change, 7*(3), 267–284.

Gates, W.L., Rowntree, P.R. and Zeng, Q.C. (1990) Validation of climate models. In J.T. Houghton, G.J. Jenkins and J.J. Ephraums (eds), *Climate Change: the IPCC Scientific Assessment*, Cambridge University Press, Cambridge, pp. 93–130.

Geller, M.A. (1989) Atmospheric circulation: variations without forcing. *Nature, 342*(6245), 15–16.

Genthon, C., Barnola, J.M., Raynaud, D., Lorius, C., Jouzel., J., Barkov, N.I., Korotkevich, Y.S. and Kotlyakov, V.M. (1987) Vostok ice core: climatic response to CO_2 and orbital forcing changes over the last climatic cycle. *Nature, 329*(6138), 414–418.

Ghan, S.J., Taylor, K.E., Penner., J.E. and Erikson, D.J. (1990) Model test of CCN-cloud albedo climate forcing. *Geophysical Research Letters, 17*(5), 607–610.

Gibbard, P.L., West R.G., Zagwijn, W.H., Balson, P.S., Burger, A.W., Funnell, B.M., Jeffery, D.H., de Jong, J., van Kolfschoten, T., Lister, A.M., Meijer, T., Norton, P.E.P., Preece, R.C., Rose, J., Stuart, A.J., Whiteman, C.A. and Zalasiewicz, J.A. (1991) Early and middle Pleistocene correlations in the southern North Sea Basin. *Quaternary Science Reviews, 10*(1), 23–52.

Gilliland, R.L. (1981) Solar radius variations over the past 265 years. *Astrophysical Journal, 248*, 1144–1155.

Giorgi, F. and Mearns, L.O. (1991) Approaches to the simulation of regional climate change — a review. *Reviews of Geophysics, 29*(2), 191–216.

Giorgi, F., Marinucci, M.R. and Visconti, G. (1990) Use of a limited-area model nested in a general circulation model for regional climate simulation over Europe. *Journal of Geophysical Research, 95*(D11), 18413–18431.

Gleick, J. (1988) *Chaos, Making a New Science*. Heinemann, London.

Gleick, P.H. (1987) Regional hydrologic consequences of increases in atmospheric CO_2 and other trace gases. *Climatic Change, 10*(2), 137–160.

Gleick, P.H. (1989) Vulnerabilities of water supply. In P. Waggoner and R.R. Revelle (eds), *Climate and Water: Climatic Variability, Climate Change, and the Planning and Management of U.S. Water Resources*, John Wiley, New York.

Godwin, Sir Harry (1975) *The History of the British Flora*. Cambridge University Press, Cambridge.

Goodess, C.M. and Palutikof, J.P. (1990) Western European regional climate scenarios in a high greenhouse gas world and agricultural impacts. In J.J. Beukema, W.J. Wolff and J.J.W.M. Brouns (eds), *Expected Effects of Climatic Change on Marine Coastal Ecosystems*, Kluwer Academic Publishers, Dordrecht, pp. 23–32.

Goodess, C.M., Palutikof, J.P. and Davies, T.D. (1988) *Studies of Climatic Effects Relevant to Site Selection and to Assessments of the Radiological Impact of Disposal at Selected Sites*. Nirex Safety Series, NSS/R137.

Goodess, C.M., Palutikof, J.P. and Davies, T.D. (1990) A first approach to assessing future climate states in the UK over very long timescales: input to studies of the integrity of radioactive waste repositories. *Climatic Change, 16*(1), 115–139.

Goodess, C.M., Palutikof, J.P. and Davies, T.D. (1992) *Studies of Climatic Effects and Impacts Relevant to Deep Underground Disposal of Radioactive Waste.* Nirex Safety Series, NSS/R267.

Gordon, A. (1982) World ocean water masses and the saltiness of the Atlantic. In *Proceedings of the Study Conference on Large-scale Oceanographic Experiments in the WCRP, Tokyo.* WMO Secretariat, Geneva.

Gordon, A. (1986) Inter-ocean exchange of thermocline water. *Journal of Geophysical Research, 91*(C4), 5037–5046.

Gordon, D., Smart, P.L., Ford, D.C., Andrews, J.N., Atkinson, T.C., Rowe, P.J. and Christopher, N.S.J. (1989) Dating of late Pleistocene interglacial and interstadial periods in the United Kingdom from speleothem growth frequency. *Quaternary Research, 31*(1), 14–26.

Gornitz, V. and Lebedeff, S. (1987) Global sea level changes during the past century. In D. Nummedal, O.H. Pilkey and J.D. Howard (eds), *Sea-level Fluctuation and Coastal Evolution,* SEPM Special Publication No. 41.

Gornitz, V. and White, T.A. (1991) The Global Coastal Hazards Data Base. In C.M. Goodess and J.P. Palutikof (eds), *Proceedings of the International Workshop on Future Climate Change and Radioactive Waste Disposal,* Nirex Safety Series, NSS/R257, pp. 214–224.

Gornitz, V., Lebedeff, S. and Hansen, J. (1982) Global sea level trend in the past century. *Science, 215,* 1611–1614.

Gough, D.O. (1990) On possible origins of relatively short-term changes in the solar structure. *Philosophical Transactions of the Royal Society of London, A, 330,* 627–640.

Gray, J.M. (1982) The last glaciers (Loch Lomond Advance) in Snowdonia, N. Wales. *Geological Journal, 17,* 111–133.

Gray, J.M. and Lowe, J.J. (eds) (1977) *Studies in the Scottish Lateglacial Environment,* Pergamon Press, Oxford.

Gregory, J.M., Jones, P.D. and Wigley, T.M.L. (1990) *Climatic Change and its Potential Effect on UK Water Resources,* Report to UK National Rivers Authority, PRS 2467-M.

Greuell, W. (1989) Glaciers and Climate: Energy Balance Studies and Numerical Modelling of the Historical Front Variations of the Hintereisferner (Austria). PhD thesis, University of Utrecht, Netherlands.

Grotch, S.L. and MacCracken, M.C. (1991) The use of general circulation models to predict regional climatic change. *Journal of Climate, 4*(3), 286–303.

Grove, J.M. (1988) *The Little Ice Age.* Methuen, London.

Guiot, J., Pons, A., de Beaulieu, J.L. and Reille, M. (1989) A 140,000-year continental climate reconstruction from two European pollen records. *Nature, 338*(6213), 309–312.

Gutenberg, B. (1941) Changes in sea level, postglacial uplift and mobility on the Earth's interior. *Bulletin of the Geological Society of America, 52,* 721–722.

Hall, A.M. and Whittington, G. (1989) Late Devensian glaciation of southern Caithness. *Scottish Journal of Geology, 25*(3), 307–324.

Hammer, C.H., Clausen, H.B. and Dansgaard, W. (1980) Greenland ice sheet evidence of post-glacial volcanism and its climatic impact. *Nature, 288*(5788), 230–235.

Hammer, C.H., Clausen, H.B. and Dansgaard, W. (1981) Past volcanism and climate revealed by Greenland ice cores. *Journal of Volcanology and Geothermal Research, 11,* 3–10.

References

Hansen, J.E., Lacis, A., Rind, D., Russell, L., Stone, P., Fung, I., Ruedy, R. and Lerner, J. (1984) Climate sensitivity: analysis of feedback mechanisms. In J.E. Hansen and T. Takahashi (eds), *Climate Processes and Climate Sensitivity*, Geophysical Monograph 29, Maurice Ewing Volume 5, American Geophysical Union, Washington, DC.

Hansen, J., Lacis, A. and Prather, M. (1989) Greenhouse effect of chlorofluorocarbons and other trace gases. *Journal of Geophysical Research, 94*(D13), 16417–16422.

Harris, C. (1991) Glacial deposits at Wylfa Head, Anglesey, North Wales: evidence for Late Devensian deposition in a non-marine environment. *Journal of Quaternary Science, 6*(1), 67–77.

Harvey, L.D.D. (1989a) An energy balance climate model study of radiative forcing and temperature response at 18 ka. *Journal of Geophysical Research, 94*(D10), 12873–12884.

Harvey, L.D.D. (1989b) Milankovitch forcing, vegetation feedback, and North Atlantic deep-water formation. *Journal of Climate, 2*(8), 800–815.

Harvey, L.D.D. (1989c) Transient climatic response to an increase of greenhouse gases. *Climatic Change, 15*(1–2), 15–30.

Hays, J.D. and Berggren, W.A. (1971) Quaternary boundaries and correlations. In B.M. Funnell and W.R. Rediel (eds), *Micropalaeontology of the Oceans*, Cambridge University Press, Cambridge, pp. 669–691.

Hays, J.D., Imbrie, J. and Shackleton, N.J. (1976) Variations in the earth's orbit: pacemaker of the ice ages. *Science, 194*, 1121–1132.

Heinrich, H. (1988) Origin and consequences of cyclic ice rafting in the north-east Atlantic Ocean during the past 130,000 years. *Quaternary Research, 29*(2), 142–152.

Heinze, C. and Maier-Reimer, E. (1989) Glacial pCO_2 reduction by the world ocean experiments with the Hamburg carbon cycle model. In *World Conference on Analysis and Evaluation of Atmospheric CO_2 Data, Present and Past*, WMO Environmental Pollution Monitoring and Research Program, Report No. 59, pp. 9–14.

Henderson-Sellers, A. (1986) Cloud changes in a warmer Europe. *Climatic Change, 8*(1), 25–52.

Henderson-Sellers, A. and McGuffie, K. (1989) Diagnosis of cloud amount increase from an analogue of a warming world. *Atmosfera, 2*, 67–101.

Henderson-Sellers, A. and Robinson, P.J. (1986) *Contemporary Climatology*, Longman Scientific and Technical, London.

Herbert, T.D. and D'Hondt, S.L. (1990) Precessional climate cyclicity in late Cretaceous–early Tertiary marine sediments: a high resolution chronometer of Cretaceous–Tertiary boundary events. *Earth and Planetary Science Letters, 99*(3), 263–275.

Herbert, T.D. and Fischer, A.G. (1986) Milankovitch climatic origin of mid-Cretaceous black shale rhythms in central Italy. *Nature, 321*(6072), 739–743.

Heusser, C.J. and Heusser, L.E. (1990) Long continental pollen sequence from Washington State (USA): correlation of upper levels with marine pollen oxygen isotope stratigraphy through substage 5e. *Palaeogeography, Palaeoclimatology, Palaeoecology, 79*(1–2), 63–71.

Heusser, C.J., Heusser, L.E. and Peteet, D.M. (1985) Late-Quaternary climatic change on the American North Pacific coast. *Nature, 315*(6019), 485–487.

Heyworth, A., Kidson, C. and Wilks, P. (1985) Late-glacial and Holocene sediments at Clarach Bay, near Aberystwyth. *Journal of Ecology, 73*, 459–480.

Hibbert, F.A. and Switsur, V.R. (1976) Radiocarbon dating of Flandrian pollen zones in Wales and northern England. *New Phytologist, 77*, 793–807.

Hillaire-Marcel, C. and Causse, C. (1989) The late Pleistocene Laurentide glacier: Th/U dating of its major fluctuations and ^{18}O range of the ice. *Quaternary Research, 32*(2), 125–138.

Hoffman, J.S., Keyes, D. and Titus, J.G. (1983) *Projecting Future Sea Level Rise: Methodology, Estimates to the Year 2100, and Research Needs.* US GPO 055-000-0236-3. Government Printing Office, Washington, DC.

Hoffman, J.S., Wells, J.B. and Titus, J.G. (1986) Future global warming and sea level rise. In G. Sigbjarnarson (ed), *Iceland Coastal and River Symposium '85,* National Energy Authority, Reykjavik.

Hood, L.L. and Jirikowic, J.L. (1990) Recurring variations of probable solar origin in the atmospheric δC^{14} time record. *Geophysical Research Letters, 17*(1), 85–88.

Horowitz, A. (1989) Continuous pollen diagrams for the last 3.5 m.y. from Israel: vegetation, climate and correlation with the oxygen isotope record. *Palaeogeography, Palaeoclimatology, Palaeoecology, 72*(1–2), 63–78.

Houghton, J.T., Jenkins, G.J. and Ephraums, J.J. (eds) (1990) *Climate Change, the IPCC Scientific Assessment.* WMO/UNEP, Cambridge University Press, Cambridge.

Hovan, S.A., Read, D.K., Pisias, N.G. and Shackleton, N.J. (1989) A direct link between the China loess and marine ^{18}O records: aeolian flux to the north Pacific. *Nature, 340*(6231), 296–298.

Hoyt, D.V. and Eddy, J.A. (1982) NCAR Technical Note TN-194 + STR. National Center for Atmospheric Research, Boulder, Colorado.

Hughes, T.J. (1985) The great Cenozoic ice sheet. *Palaeogeography, Palaeoclimatology, Palaeoecology, 50,* 9–44.

Hughes, T.J. (1987) Deluge II and the continent of doom: rising sea level and collapsing Antarctic ice. *Boreas, 16*(2), 89–100.

Huntley, B. (1990) European vegetation history: palaeovegetation maps from pollen data — 13000 yr BP to present. *Journal of Quaternary Science, 5*(2), 103–122.

Huntley, B. and Birks, H.J.B. (1983) *An Atlas of Past and Present Pollen Maps for Europe, 0–13,000 years ago.* Cambridge University Press, Cambridge.

Huntley, B. and Prentice, I.C. (1988) July temperatures in Europe from pollen data, 6000 years before present. *Science, 241,* 687–689.

Huybrechts, P. and Oerlemans, J. (1990) Response of the Antarctic ice sheet to future greenhouse warming. *Climate Dynamics, 5,* 93–102.

Huybrechts, P., Letréguilly, A. and Reeh, N. (1991) The Greenland ice sheet and greenhouse warming. *Palaeogeography, Palaeoclimatology, Palaeoecology (Global and Planetary Change Section), 89,* 399–412.

Hyde, W.T. and Peltier, W.R. (1985) Sensitivity experiments with a model of the ice age cycle: the response to harmonic forcing. *Journal of the Atmospheric Sciences, 42*(20), 2170–2188.

Hyde, W.T. and Peltier, W.R. (1987) Sensitivity experiments with a model of the ice age cycle: the response to Milankovitch forcing. *Journal of the Atmospheric Sciences, 44*(10), 1351–1374.

Hyde, W.T., Crowley, T.J., Kim, K-Y. and North, G.R. (1989) Comparison of GCM and energy balance model simulations of seasonal temperature changes over the past 18 000 years. *Journal of Climate, 2*(8), 864–887.

Imbrie, J. (1985) A theoretical framework for the Pleistocene ice ages. *Journal of the Geological Society, 142*(3), 417–432.

Imbrie, J. and Imbrie, J.Z. (1980) Modeling the climatic response to orbital variations. *Science, 207,* 943–953.

Imbrie, J. and Imbrie, K.P. (1979) *Ice Ages, Solving the Mystery.* Macmillan Press, London.

Imbrie, J., Hays, J.D., Martinson, D.G., McIntyre, A., Mix, A.C., Morley,

References

J.J., Pisias, N.G., Prell, W.L. and Shackleton, N.J. (1984) The orbital theory of Pleistocene climate: support from a revised chronology of the marine ^{18}O record. In A. Berger, J. Imbrie, J. Hays, G. Kukla and B. Saltzman (eds), *Milankovitch and Climate*, D. Reidel, Dordrecht, pp. 269–306.

Imbrie, J., McIntyre, A. and Mix, A. (1989) Oceanic response to orbital forcing in the Late Quaternary: observational and experimental strategies. In A. Berger, S. Schneider and J.C. Duplessy (eds), *Climate and Geo-sciences*, Kluwer Academic Publishers, Dordrecht, pp. 121–164.

Ingram, W.J., Wilson, C.A. and Mitchell, J.F.B. (1989) Modelling climate change: an assessment of sea ice and surface albedo feedbacks. *Journal of Geophysical Research, 94*(D6), 8609–8622.

Iverson, J. (1958) The bearing of glacial and interglacial epochs on the formation and extinction of plant taxa. *Uppsala Univ. Arsskr., 6*, 210.

Jäger, J. (1988) *Developing Policies for Responding to Climatic Change. A Summary of the Discussions and Recommendations of the Workshops Held in Villach (28 September–2 October 1987) and Bellagio (9–13 November 1987), under the Auspices of the Beijer Institute*, World Climate Programme Impact Studies, WCIP-1, WMO-TD-No.225, World Meteorological Organization/United Nations Environment Programme, WMO, Geneva.

Jäger, J. and Ferguson, H.L. (eds) (1991) *Climate Change: Science, Impacts and Policy. Proceedings of the Second World Climate Conference*, Cambridge University Press, Cambridge.

James, I.N. and James, P.M. (1989) Ultra-low-frequency variability in a simple atmospheric circulation model. *Nature, 342*(6245), 53–55.

Jansen, E. and Veum, T. (1990) Evidence for two-step deglaciation and its impact on North Atlantic deep-water circulation. *Nature, 343*(6259), 612–616.

Jardine, W.G., Dickson, J.H., Haughton, P.D.W, Harkness, D.D., Bowen, D.Q. and Sykes, G.A. (1988) A late Middle Devensian interstadial site at Sourlie, near Irvine, Strathclyde. *Scottish Journal of Geology, 24*(3), 288–295.

Jelgersma, S. (1961) Holocene sea-level changes in the Netherlands. *Mededelingen van de Geologische Stichting*, Series C, 6(7), 1–101.

Jensen, K.A. and Knudsen, K.L. (1988) Quaternary foraminiferal stratigraphy in boring 81/29 from the central North Sea. *Boreas, 17*, 274–287.

Johnson, R.H. (1975) Some late Pleistocene involutions at Dalton-in-Furness, northern England. *Geological Journal, 10*, 23–33.

Jones, G.A. and Keigwin, L.D. (1988) Evidence from Fram Strait (78°N) for early deglaciation. *Nature, 336*(6194), 56–59.

Jones, G.A. and Ruddiman, W.F. (1982) Assessing the global meltwater spike. *Quaternary Research, 17*(2), 148–172.

Jones, P.D. (1990) The climate of the last thousand years. *La Recherche, 21*(219), 304–313.

Jones, P.D., Raper, S.C.B., Santer, B., Cherry, B.S.G., Goodess, C.M., Kelly, P.M., Wigley, T.M.L., Bradley, R.S. and Diaz, H.F. (1985) *A Grid Point Surface Air Temperature Data Set for the Northern Hemisphere*, DOE/EV/10098-2, Carbon Dioxide Research Division, Office of Basic Energy Sciences, Office of Energy Research, US Department of Energy, Washington, DC.

Jones, P.D., Wigley, T.M.L. and Wright, P.B. (1986) Global temperature variations between 1861 and 1984. *Nature, 322*(6078), 430–434.

Jong, J. (1988) Climatic variability during the past three million years, as indicated by vegetational evolution in northwest Europe and with emphasis on data from the Netherlands. *Philosophical Transactions of the Royal Society of London B, 318*, 603–617.

Jouzel, J., Lorius, C., Petit, J.R., Genthon, C., Barkov, N.I., Kotlyakov, V.M. and

Petrov, V.M. (1987) Vostock ice core: a continuous isotope temperature record over the last climatic cycle (160,000 years). *Nature, 329*(6138), 403–408.

Jouzel, J., Raisbeck, G., Benoist, J.P., Yiou, F., Lorius, C., Raynaud, D., Petit, J.R., Barkov, N.I., Korotkevitch, Y.S. and Kotlyakov, V.M. (1989) A comparison of deep Antarctic ice cores and their implications for climate between 65,000 and 15,000 years ago. *Quaternary Research, 31*(2), 135–150.

Karl, T.R., Tarpley, J.D., Quayle, R.G., Diaz, H.F., Robinson, D.A. and Bradley, R.S. (1989) The recent climate record — what it can and cannot tell us. *Reviews of Geophysics, 27*(3), 405–430.

Karl, T.R., Wang, W.C., Schlesinger, M.E., Knight, R.W. and Portman, D. (1990) A method of relating general circulation model simulated climate to the observed local climate. Part 1: seasonal statistics. *Journal of Climate, 3,* 1053–1079.

Kashiwaya, K., Atkinson, T.C. and Smart, P.L. (1991) Periodic variations in late Pleistocene speleothem abundance in Britain. *Quaternary Research, 35*(2), 190–196.

Keatinge, T.H. and Dickson, J.H. (1979) Mid-Flandrian changes in vegetation on mainland Orkney. *New Phytologist, 82,* 585–612.

Keffer, T., Martinson, D.G. and Corliss, B.H. (1988) The position of the Gulf Stream during Quaternary glaciations. *Science, 241,* 440–442.

Keigwin, L.D. and Boyle, E.A. (1989) Late Quaternary paleochemistry of high-latitude surface waters. *Palaeogeography, Palaeoclimatology, Palaeoecology, 73*(1–2), 85–106.

Keigwin, L.D. and Jones, G.A. (1989) Glacial-Holocene stratigraphy, chronology, and paleoceanographic observations on some North Atlantic sediment drifts. *Deep-Sea Research, Part A Oceanographic Research Papers, 36*(6), 845–868.

Kelly, P.M. and Sear, C.B. (1984) Climatic impact of explosive volcanic eruptions. *Nature, 311*(5988), 740–743.

Kerney, M.J. (1963) Late-glacial deposits on the chalk of south-east England. *Philosophical Transactions of the Royal Society of London B, 246,* 203–254.

Kerney, M.J., Brown, E.H. and Chandler, T.J. (1964) The late-glacial and post-glacial history of the chalk escarpment near Brook, Kent. *Philosophical Transactions of the Royal Society of London B, 248,* 135–204.

Kerr, R.A. (1988) Sunspot-weather link holding up. *Science, 242,* 1124–1125.

Kerr, R.A. (1989a) Did the roof of the world start an ice age? *Science, 244,* 1441–1442.

Kerr, R.A. (1989b) Volcanoes can muddle the greenhouse. *Science, 245,* 127–128.

Kerr, R.A. (1989c) Bringing down the sea level rise. *Science, 246,* 1563.

Kiehl, J.T. and Ramanathan, V. (1990) Comparison of cloud forcing derived from the Earth Radiation Budget Experiment with that simulated by the NCAR Community Climate Model. *Journal of Geophysical Research, 95*(D8), 11679–11698.

Kim, J.-W., Chang, J.-T., Baker, N.L., Wilks, D.S. and Gates, W.L. (1984) The statistical problem of climate inversion: determination of the relationship between local and large-scale climate. *Monthly Weather Review, 112*(10), 2069–2077.

Klige, R.K. (1982) Oceanic level fluctuations in the history of the earth. In *Sea and Oceanic Level Fluctuations for 15,000 Years,* Academy of Sciences of the USSR, Institute of Geography, Nauka Publishing House, Moscow, pp. 11–22.

Koerner, R.M. (1989) Ice core evidence for extensive melting of the Greenland ice sheet in the last interglacial. *Science, 244,* 964–968.

Kolstrup, E. (1980) Climate and stratigraphy in northwestern Europe between 30,000 B.P. and 13,000 B.P., with special reference to the Netherlands. *Mededelingen Rijks Geologische Dienst, 32*(15), 181–253.

Ku, T.-L., Ivanovich, M. and Luo, S. (1990) U-series dating of last interglacial high sea stands: Barbados revisited. *Quaternary Research, 33*(2), 129–147.

Kudrass, H.R., Erlenkeuser, H., Vollbrecht, R. and Weiss, W. (1991) Global nature of the Younger Dryas cooling event inferred from oxygen isotope data from Sulu Sea cores. *Nature, 349*(6308), 406–409.

Kuenen, P.H. (1950) *Marine Geology.* Wiley, New York.

Kuhle, M. (1987) Subtropical mountain- and highland-glaciation as ice age triggers and the warming of the glacial periods in the Pleistocene. *Geojournal, 14*(4), 393–421.

Kuhn, J.R., Libbrecht, K.G. and Dicke, R.H. (1988) The surface temperature of the sun and changes in the solar constant. *Science, 242,* 908–910.

Kuhn, M. (1992) Contribution of glaciers to sea level rise. In R.A. Warrick, E.M. Barrow and T.M.L. Wigley (eds), *Climate and Sea Level Change: Observations, Projections and Implications,* Cambridge University Press, Cambridge, in press.

Kukal, Z. (1990) The rates of geological processes. *Earth Science Reviews, 28*(1–2), 5–284.

Kukla, G. (1989) Long continental records of climate — an introduction. *Palaeogeography, Palaeoclimatology, Palaeoecology, 72,* 1–9.

Kukla, G. and An, Z. (1989) Loess stratigraphy in central China. *Palaeogeography, Palaeoclimatology, Palaeoecology, 72,* 203–225.

Kukla, G. and Kukla, H.J. (1972) Insolation regime of interglacials. *Quaternary Research, 2*(3), 412–424.

Kukla, G., Berger, A., Lotti, R. and Brown, J. (1981) Orbital signature of interglacials. *Nature, 290*(5804), 295–300.

Kukla, G., Heller, F., Ming, L.X., Chun, X.T., Sheng, L.T. and Sheng, A.Z. (1988) Pleistocene climates in China dated by magnetic susceptibility. *Geology, 16,* 811–814.

Kutzbach, J.E. and Guetter, P.J. (1984) Sensitivity of late-glacial and Holocene climates to the combined effects of orbital parameter changes and lower boundary condition changes: 'snapshot' simulations with a general circulation model for 18, 9 and 6 ka BP. *Annals of Glaciology, 5,* 85–87.

Kutzbach, J.E. and Guetter, P.J. (1986) The influence of changing orbital parameters and surface boundary conditions on climate simulations for the past 18000 years. *Journal of the Atmospheric Sciences, 43*(16), 1726–1759.

Kutzbach, J.E. and Street-Perrott, F.A. (1985) Milankovitch forcing of fluctuations in the level of tropical lakes from 18 to 0 kyr B.P. *Nature, 317*(6033), 130–134.

Kutzbach, J.E. and Wright, H.E., Jr. (1985) Simulation of the climate of 18,000 years BP: results for the North American/North Atlantic/European sector and comparison with the geologic record of North America. *Quaternary Science Reviews, 4,* 147–187.

Kutzbach, J.E., Guetter, P.J., Ruddiman, W.F. and Prell, W.L. (1989) Sensitivity of climate to Late Cenozoic uplift in southern Asia and the American west: numerical experiments. *Journal of Geophysical Research, 94* (D15), 18393–18408.

Labeyrie, L.D., Pichon, J.J., Labracherie, M., Ippolito, P., Duprat, J. and Duplessy, J.C. (1986) Melting history of Antarctica during the past 60,000 years. *Nature, 322*(6081), 701–706.

Labeyrie, L.D., Duplessy, J.C. and Blanc, P.L. (1987) Variations in the mode of formation and temperature of oceanic deep waters over the past 125,000 years. *Nature, 327*(6122), 477–481.

Labitzke, K. and van Loon, H. (1988) Associations between the 11-year solar cycle, the QBO and the atmosphere. Part I: The troposphere and stratosphere in the Northern Hemisphere in winter. *Journal of Atmospheric and Terrestrial Physics, 50,* 197–206.

Labitzke, K. and van Loon, H. (1989a) Association between the 11-year solar cycle, the QBO, and the atmosphere. Part III: Aspects of the association. *Journal of Climate, 2*(6), 554–565.

Labitzke, K. and van Loon, H. (1989b) The 11-year solar cycle in the stratosphere in the northern summer. *Annales Geophysicae, 7,* 595–598.

Labitzke, K. and van Loon, H. (1990) Associations between the 11-year solar cycle, the quasi-biennial oscillation and the atmosphere: a summary of recent work. *Philosophical Transactions of the Royal Society of London A, 330,* 577–590.

LaMarche, V.C. and Hirschboeck, K.K. (1984) Frost rings in trees as records of major volcanic eruptions. *Nature, 307*(5947), 121–126.

Lamb, H.H. (1970) Volcanic dust in the atmosphere; with a chronology and assessment of its meteorological significance. *Philosophical Transactions of the Royal Society A, 266,* 452–533.

Lamb, H.H. (1975) Data given in Appendix A: Survey of past climates. In *Understanding Climatic Change, a Program for Action.* National Academy of Sciences, Washington, DC, pp 127–195.

Lamb, H.H. (1977a) *Climate, Present, Past and Future,* 2 vols, Methuen, London.

Lamb, H.H. (1977b) The late Quaternary history of the climate of the British Isles. In F.W. Shotton (ed.), *British Quaternary Studies,* Clarendon Press, Oxford, pp. 283–298.

Lamb, H.H. (1979) Climatic variation and changes in wind and ocean circulation: the Little Ice Age in the Northeast Atlantic. *Quaternary Research, 11*(1), 1–20.

Lamb, H.H. and Johnson, A.I. (1966) *Secular Variations of the Atmospheric Circulation since 1750,* Geophysical Memoir 110, London.

Lamb, P.J. (1987) On the development of regional climatic scenarios for policy-orientated climatic-impact assessment. *Bulletin of the American Meteorological Society, 68*(9), 1116–1123.

Lambeck, K., Johnston, P. and Nakada, M. (1990) Holocene glacial rebound and sea-level change in NW Europe. *Geophysical Journal International, 103*(2), 451–468.

Lashof, D.A. (1989) The dynamic greenhouse: feedback processes that may influence future concentrations of atmospheric trace gases and climatic change. *Climatic Change, 14*(3), 213–242.

Lawson, T.J. (1987) Loch Lomond advance glaciers in Assynt, Sutherland, and their palaeoclimatic implications. *Scottish Journal of Geology, 22*(3), 289–298.

Lean, J. and Foukal, P. (1988) A model of solar luminosity modulation by magnetic activity between 1954 and 1984. *Science, 240,* 906–907.

Lean, J. and Warrilow, D.A. (1989) Simulation of the regional climatic impact of Amazon deforestation. *Nature, 342*(6248), 411–413.

Legrand, M.R., Delmas, R.J. and Charlson, R.J. (1988a) Climate forcing implications from Vostok ice-core sulphate data. *Nature, 334*(6181), 418–420.

Legrand, M.R., Lorius, C., Barkov, N.I. and Petrov, V.N. (1988b) Vostok (Antarctica) ice core: atmospheric chemistry changes over the last climatic cycle (160,000 years). *Atmospheric Environment, 22*(2), 317–332.

Legrand, M.R, Feniet-Saigne, C., Saltzman, E.S., Germain, C., Barkov, N.I. and Petrov, V.N. (1991) Ice-core record of oceanic emissions of dimethylsulphide during the last climate cycle. *Nature, 350*(6314), 144–146.

Lehman, S.J., Jones, G.A., Keigwin, L.D., Andersen, E.S., Butenko, G. and Østmo, S.R. (1991) Initiation of Fennoscandian ice-sheet retreat during the last deglaciation. *Nature, 349*(6309), 513–516.

Le Roy Ladurie, E. (1967) *Histoire du climat depuis l'an mil,* Flammarion, Paris. (A translation by B. Bray, published by Doubleday and Co., appeared in 1971.)

Letréguilly, A., Huybrechts, P. and Reeh, N. (1991) Steady-state characteristics of the Greenland ice sheet under different climates. *Journal of Glaciology, 37*(125), 149–157.

References

Le Treut, H. and Ghil, M. (1983) Orbital forcing, climatic interactions and glaciation cycles. *Journal of Geophysical Research, 88*(C9), 5167–5190.

Lezine, A.M. and Casanova, J. (1991) Correlated oceanic and continental records demonstrate past climate and hydrology of North Africa (0–140 ka). *Geology, 19*(4), 307–310.

Lezine, A.M. and Hooghiemstra, H. (1990) Land–sea comparisons during the last glacial–interglacial transition: pollen records from west Tropical Africa. *Palaeogeography, Palaeoclimatology, Palaeoecology, 79*(3–4), 313–331.

Li, W.X., Lundberg, J., Dickin, A.P., Ford, D.C., Schwarcz, H.P., McNutt, R. and Williams, D. (1989) High-precision mass-spectrometric uranium-series dating of cave deposits and implications for paleoclimate studies. *Nature, 339*(6225), 534–536.

Lindstrom, D.R. and MacAyeal, D.R. (1989) Scandinavian, Siberian, and Arctic Ocean glaciation: effect of Holocene atmospheric CO_2 variations. *Science, 245*, 628–631.

Lindzen, R.S. (1990) Some coolness concerning global warming. *Bulletin of the American Meteorological Society, 71*(3), 288–299.

Lingle, C.S. (1985) A model of a polar ice stream and future sea level rise due to possible drastic retreat of the West Antarctic Ice Sheet. In *Glaciers, Ice Sheets and Sea Level: Effects of a CO_2-induced Climatic Change,* National Academy Press, Washington, DC, pp. 317–330.

Lisitzin, E. (1974) *Sea-level Changes.* Elsevier Oceanography Series 8, Elsevier, Amsterdam.

Lockwood, J.G. (1985) *World Climatic Systems.* Edward Arnold, London.

Lorenz, E.N. (1968) Climatic determinism. Meteorological Monographs, *8*, 1–3.

Lorenz, E.N. (1976) Nondeterministic theories of climatic change. *Quaternary Research, 6*(4), 495–506.

Lorenz, E.N. (1991) Chaos, spontaneous climatic variations and detection of the greenhouse effect. In M.E. Schlesinger (ed.), *Greenhouse-Gas-Induced Climatic Change: a Critical Appraisal of Simulations and Observations*, Developments in Atmospheric Science 19, Elsevier, Amsterdam, pp. 445–453.

Lorius, C., Jouzel, J., Raynaud, D., Hansen, J. and Le Treut, H. (1990) The ice-core record: climate sensitivity and future greenhouse warming. *Nature, 347*(6289), 139–145.

Loubere, P. and Moss, K. (1986) Late Pliocene climatic change and the onset of Northern Hemisphere glaciation as recorded in the northeast Atlantic Ocean. *Geological Society of America Bulletin, 97*(7), 818–828.

Lough, J.M. and Fritts, H.C. (1987) An assessment of the possible effects of volcanic eruptions on North American climate using tree ring data, 1602 to 1900 A.D. *Climatic Change, 10*(3), 219–240.

Lough, J.M., Wigley, T.M.L. and Palutikof, J.P. (1983) Climate and climate impact scenarios for Europe in a warmer world. *Journal of Climate and Applied Meteorology, 22*(10), 1673–1684.

Lowe, S. (1981) Radiocarbon dating and stratigraphic resolution in Welsh lateglacial chronology. *Nature, 293*(5289), 210–212.

Maasch, K.A. and Saltzman, B. (1990) A low-order dynamical model of global climatic variability over the full Pleistocene. *Journal of Geophysical Research, 95*(D2), 1955–1963.

MacCracken, M.C. and Luther, F.M. (eds) (1985) *Projecting the Climatic Effects of Increasing Carbon Dioxide*, DOE/ER-0237, Carbon Dioxide Research Division, Office of Energy Research, US Department of Energy, Washington, DC.

Manabe, S. and Broccoli, A.J. (1985) The influence of continental ice sheets on the climate of an ice age. *Journal of Geophysical Research, 90*(D1), 2167–2190.

Manabe, S. and Stouffer, R.J. (1988) Two stable equilibria of a coupled ocean–atmosphere model. *Journal of Climate, 1*(9), 841–866.

Manabe, S. and Wetherald, R.T. (1987) Large-scale changes of soil wetness induced by an increase in atmospheric carbon dioxide. *Journal of the Atmospheric Sciences, 44*(8), 1211–1235.

Manabe, S., Bryan, K. and Spelman, M.J. (1990) Transient response of a global ocean–atmosphere model to a doubling of atmospheric carbon dioxide. *Journal of Physical Oceanography, 20*(5), 722–749.

Manabe, S., Stouffer, R.J., Spelman, M.J. and Bryan, K. (1991) Transient responses of a coupled ocean–atmosphere model to gradual changes of atmospheric CO_2. Part 1: annual mean response. *Journal of Climate, 4*(8), 785–818.

Manley, G. (1975) Fluctuations and persistence of snow cover in marginal oceanic climates. In *Proceedings of WMO/IAMAP Symposium on Long-term Climatic Fluctuations, Norwich, 18–23 August 1975*. WMO No.421, WMO, Geneva, pp. 183–188.

Martinson, D.G., Pisias, N.G., Hays, J.D., Imbrie, J., Moore, T.C. and Shackleton, N.J. (1987) Age dating and the orbital theory of the ice ages: development of a high-resolution 0 to 300,000 year chronostratigraphy. *Quaternary Research, 27*(1), 1–29.

Mass, C.F. and Portman, D.A. (1989) Major volcanic eruptions and climate: a critical evaluation. *Journal of Climate, 2*(6), 566.

Mazzarella, A. and Palumbo, A. (1989) The 11-year ozone modulation of extreme surface air temperatures *Theoretical and Applied Climatology, 40*(3), 155–160.

McIntyre, A., Kipp, N.G., Bé, A.W.H., Crowley, T., Kellogg, T.B., Gardner, J.V., Prell, W. and Ruddiman, W.F. (1976) Glacial North Atlantic 18,000 years ago: a CLIMAP reconstruction. In *Investigation of Late Quaternary Paleoceanography and Paleoclimatology*, Geological Society of America Memoir 145, pp. 43–76.

Mearns, L.O., Schneider, S.H., Thompson, S.L. and McDaniel, L.R. (1990) Analysis of climate variability in general circulation models — comparison with observations and changes in variability in $2 \times CO_2$ experiments. *Journal of Geophysical Research, 95*(D12), 20469–20490.

Meehl, G.A. and Washington, W.A. (1988) A comparison of soil-moisture sensitivity in two global climate models. *Journal of the Atmospheric Sciences, 45*(9), 1476–1492.

Meier, M.F. (1984) Contribution of small glaciers to global sea level. *Science, 226,* 1418–1420.

Meier, M.F. (1990) Greenhouse effect: reduced rise in sea level. *Nature, 343*(6254), 115–116.

Mercer, J.H. (1978) West Antarctic ice sheet and CO_2 greenhouse effect: a threat of disaster. *Nature, 271*(5643), 321–325.

Mesolella, K.J., Matthews, R.K., Broecker, W.S. and Thurber, D.L. (1969) The astronomical theory of climatic change: Barbados data. *Journal of Geology, 77,* 250–274.

Milankovitch, M.M. (1941) *Canon of Insolation and the Ice Age Problem.* Königlich Serbische Akademie, Belgrade. English translation by the Israel Program for Scientific Translations, United States Department of Commerce and the National Science Foundation, Washington, DC.

Mitchell, G.F., Penny, L.F., Shotton, F.W. and West, R.G. (1973) *A Correlation of Quaternary Deposits in the British Isles.* Geological Society of London, Special Report 4.

Mitchell, J.F.B. and Warrilow, D.A. (1987) Summer dryness in northern

References

midlatitudes due to increased CO_2. *Nature, 330*(6145), 238–240.

Mitchell, J.F.B., Grahame, N.S. and Needham, K.J. (1988) Climate simulations for 9000 years before present: seasonal variations and effect of the Laurentide ice sheet. *Journal of Geophysical Research, 93*(D7), 8283–8303.

Mitchell, J.F.B., Senior, C.A. and Ingram, W.J. (1989) CO_2 and climate: a missing feedback? *Nature, 341*(6238), 132–134.

Mitchell, J.F.B., Manabe, S., Tokioka, T. and Meleshko, V. (1990) Equilibrium climate change. In J.T. Houghton, G.J. Jenkins and J.J. Ephraums (eds), *Climate Change, the IPCC Scientific Assessment*, WMO/UNEP, Cambridge University Press, Cambridge, pp. 131–172.

Mitchell, J.M. (1976) An overview of climatic variability and its causal mechanisms. *Quaternary Research, 6*(4), 481–494.

Mitchell, J.M. (1977) The changing climate. In *Energy and Climate, Studies in Geophysics*, National Academy of Sciences, Washington, DC, pp. 51–58.

Mitchell, J.M., Stockton, C.W. and Meko, D.M. (1979) Evidence of a 22 year rhythm of drought in the western United States related to the Hale solar cycle since the 17th century. In B.M. McCormack and T.A. Seliga (eds), *Solar-terrestrial Influences on Weather and Climate*, D. Reidel, Dordrecht, pp. 125–144.

Mix, A.C. (1987) Hundred-kiloyear cycle queried. *Nature, 327*(6121), 370.

Mix, A.C. (1989) Influence of productivity variations on long-term atmospheric CO_2. *Nature, 337*(6207), 541–544.

Mix, A.C. and Pisias, N.G. (1988) Oxygen isotope analyses and deep-sea temperature changes: implications for rates of oceanic mixing. *Nature, 331*(6153), 249–251.

Mix, A.C. and Ruddiman, W.F. (1984) Oxygen-isotope analyses and Pleistocene ice volumes. *Quaternary Research, 21*, 1–20.

Molnar, P. and England, P. (1990) Late Cenozoic uplift of mountain ranges and global climate change: chicken or egg? *Nature, 346*(6279), 29–34.

Monaghan, M.C. and Holdsworth, G. (1990) The origin of non-sea-salt sulphate in the Mount Logan ice core. *Nature, 343*(6255), 245–248.

Monteith, J.L. (1981) Climatic variation and the growth of crops. *Quarterly Journal of the Royal Meteorological Society, 107*(454), 749–774.

Mörner, N.-A. (1969) The late Quaternary history of the Kattegatt Sea and the Swedish west coast: deglaciation, shorelevel displacement, chronology, isostasy and eustasy. *Sveriges Geologiske Undersökning*, Series C, 640, 1–487.

Morrison, L.V., Stephenson, F.R. and Parkinson, J. (1988) Diameter of the sun in AD 1715. *Nature, 331*(6155), 421–423.

Nakada, M. and Lambeck, K. (1988) The melting history of the late Pleistocene Antarctic ice sheet. *Nature, 333*(6168), 36–40.

Neeman, B.U., Joseph, J.H. and Ohring, G. (1988a) A vertically integrated snow/ice model over land/sea for climate models. 1. Development. *Journal of Geophysical Research, 93*(D4), 3663–3675.

Neeman, B.U., Joseph, J.H. and Ohring, G. (1988b) A vertically integrated snow/ice model over land/sea for climate models. 2. Impact on orbital change experiments. *Journal of Geophysical Research, 93*(D4), 3677–3695.

Neftel, A., Oeschger, H., Schwander, J., Stauffer, B. and Zumbrunn, R. (1982) Ice core sample measurements give atmospheric CO_2 content during the past 40,000 yr. *Nature, 295*(5846), 220–223.

Neftel, A., Oeschger, H., Staffelbach, T. and Stauffer, B. (1988) CO_2 record in the Byrd ice core 50,000–5,000 years BP. *Nature, 331*(6157), 609–611.

Newell, C.G. and Self, S. (1982) The volcanic explosivity index (VEI): an estimate of explosive magnitude for historical volcanism. *Journal of Geophysical Research, 87*(C2), 1231–1238.

Newell, N.E., Newell, R.E., Hsiung, J. and Zhongxiang, W. (1989) Global marine temperature variation and the solar magnetic cycle. *Geophysical Research Letters, 16*(4), 311–314.

Nicolis, C. and Nicolis, G. (1984) Is there a climate attractor? *Nature, 311,*(5986) 529–532.

Oerlemans, J. (1980) Model experiments on the 100,000 year glacial cycle. *Nature, 287*(5781), 430–432.

Oerlemans, J. (1982) Glacial cycles and ice-sheet modelling. *Climatic Change, 4*(4), 353–374.

Oerlemans, J. (1989) A projection of future sea level. *Climatic Change, 15*(1–2), 151–174.

Oerlemans, J. (1992) A model for the surface balance of ice masses: Part I, Alpine glaciers. *Zeitschrift für Gletscherkunde und Glazialgeologie,* submitted.

Oerlemans, J. and Van Der Veen, C.J. (1984) *Ice Sheets and Climate.* D. Reidel, Dordrecht.

Olsen, P.E. (1986) A 40-million-year lake record of early Mesozoic orbital climatic forcing. *Science, 234,* 842–848.

Overpeck, J.T., Peterson, L.C., Kipp, N., Imbrie, J. and Rind, D. (1989) Climate change in the circum-North Atlantic region during the last deglaciation. *Nature, 338*(6216), 553–557.

Padmanabhan, G. and Rao, A.R. (1990) On the crosscorrelation between drought indices and solar activity. *Theoretical and Applied Climatology, 41*(1–2), 55–61.

Palmer, T. (1989) A weather eye on unpredictability. *New Scientist, 124*(1690), 56–59.

Palutikof, J.P., Wigley, T.M.L. and Lough, J.M. (1984) *Seasonal Climate Scenarios for Europe and North America in a High-CO$_2$, Warmer World.* US Department of Energy, Carbon Dioxide Research Division, Technical Report, TR012 Washington, DC.

Palutikof, J.P., Guo, X. and Barthelmie, R.J. (1990) Wind energy and the greenhouse effect. In T.D. Davies, J.A. Halliday and J.P. Palutikof (eds), *Wind Energy Conversion 1990,* Mechanical Engineering Publications, London, pp. 123–128.

Payne, A.J., Sugden, D.E. and Clapperton, C.M. (1989) Modeling the growth and decay of the Antarctic peninsula ice sheet. *Quaternary Research, 31*(2), 119–134.

Peacock, J.D., Harkness, D.D., Housley, R.A., Little, J.A. and Paul, M.A. (1989) Radiocarbon ages for a glaciomarine bed associated with the maximum of the Loch Lomond Readvance in west Benderloch Argyll. *Scottish Journal of Geology, 25*(1), 69–79.

Peglar, S. (1979) A radiocarbon-dated pollen diagram from Loch of Winless, Caithness, north-east Scotland. *New Phytologist, 82,* 245–263.

Peltier, W.R. (1987) A relaxation oscillator model of the ice age cycle. In *Irreversible Phenomena and Dynamical Systems Analysis in Geosciences,* NATO Advanced Study Institutes Series, Series C, Mathematical and Physical Sciences, D. Reidel, Dordrecht, Vol. 192, pp. 399–416.

Peltier, W.R. (1988a) Lithospheric thickness, Antarctic deglaciation history, and ocean basin discretization effects in a global model of postglacial sea level change: a summary of some sources of nonuniqueness. *Quaternary Research, 29*(2), 93–112.

Peltier, W.R. (1988b) Global sea level and Earth rotation. *Science, 240,* 895–901.

Peltier, W.R. and Hyde, W. (1984) A model of the ice age cycle. In A. Berger, J. Imbrie, J. Hays, G. Kukla and B. Saltzman (eds), *Milankovitch and Climate,* D. Reidel, Dordrecht, pp. 565–580.

Peltier, W.R. and Tushingham, A.M. (1989) Global sea level rise and the greenhouse effect: might they be connected? *Science, 244,* 806–810.

References

Peltier, W.R. and Tushingham, A.M. (1991) Influence of glacial isostatic adjustment on tide gauge measurements of secular sea level. *Journal of Geophysical Research*, *96*(B4), 6779–6796.

Pennington, W. (1965) Pollen analysis at a microlithic site at Drigg. *Transactions of the Cumbrian Western Antiquities and Archaeological Society*, *65*, 82–85.

Pennington, W. (1970) Vegetational history in the north-west of England: a regional synthesis. In D. Walker and R.G. West (eds), *Studies in the Vegetational History of the British Isles*, Cambridge University Press, Cambridge, pp. 41–79.

Pennington, W. (1978) Quaternary geology. In F. Moseley (ed.), *The Geology of the Lake District*, Yorkshire Geological Society, Leeds, pp. 207–225.

Perry, A.H. and Walker, J.M. (1977) *The Ocean–Atmosphere System*. Longman, London.

Pestiaux, P., Van Der Mersch, I., Berger, A. and Duplessy, J.C. (1988) Paleoclimatic variability at frequencies ranging from 1 cycle per 10000 years to 1 cycle per 1000 years: evidence for nonlinear behaviour of the climate system. *Climatic Change*, *12*(1), 9–37.

Peteet, D.M., Vogel, J.S., Nelson, D.E., Southon, J.R., Nickmann, R.J. and Heusser, L.E. (1990) Younger Dryas climatic reversal in northeastern USA? AMS ages for an old problem. *Quaternary Research*, *33*(2), 219–230.

Peterson, G.M. (1983) Recent pollen spectra and zonal vegetation in the western U.S.S.R. *Quaternary Science Reviews*, *2*, 281–321.

Petit, J.R., Mounier, L., Jouzel, J., Korotkevich, Y.S., Kotlyakov, V.I. and Lorius, C. (1990) Palaeoclimatological and chronological implications of the Vostok core dust record. *Nature*, *343*(6253), 56–58.

Péwé, T.L. (1966) Palaeoclimatic significance of fossil ice wedges. *Biuletyn Peryglacjalny*, *15*, 65–73.

Pinter, N. and Gardner, T.W. (1989) Construction of a polynomial model of glacio-eustatic fluctuation — estimating paleo-sea levels continuously through time. *Geology*, *17*(4), 295–298.

Pisias, N.G. and Shackleton, N.J. (1984) Modelling the global climate response to orbital forcing and atmospheric carbon dioxide. *Nature*, *310*(5980), 757–759.

Pitman, A.J., Henderson-Sellers, A. and Yang, Z.L. (1990) Sensitivity of regional climates to localized precipitation in global models. *Nature*, *346*(6286), 734–737.

Pittock, A.B. (1978) Long-term sun–weather relations. *Review of Geophysical and Space Physics*, *16*, 400–420.

Pitty, A.F. (1992) *Aspects of Glacial Erosion and Deposition, with Particular Reference to West Cumbria and North Caithness*. Nirex Safety Series, in preparation.

Pitty, A.F., Stone, S.J. and Clayton, K.M. (1992) *Neotectonics — an Enquiry into the Structural Stability of the British Isles in Latest Geological Times and at the Present Day*. Nirex Safety Series, NSS/R236, in preparation.

Platt, C.M.R. (1989) The role of cloud microphysics in high-cloud feedback effects on climate change. *Nature*, *341*(6241), 428–429.

Polar Research Board (1985) *Glaciers, Ice Sheets and Sea Level: Effect of a CO_2-induced Climatic Change*, Report of a Workshop held in Seattle, Washington, September 13–15, 1984. US DOE/ER/60235-1.

Pollard, D. (1982) A simple ice sheet model yields a realistic 100 kyr glacial cycle. *Nature*, *296*(5855), 334–338.

Pollard, D. (1983a) A coupled climate ice sheet model applied to the Quaternary ice ages. *Journal of Geophysical Research*, *88*(C12), 7705–7718.

Pollard, D. (1983b) Ice age simulations with a calving ice-sheet model. *Quaternary Research*, *20*(1), 30–48.

Pollard, D. (1984) Some ice-age aspects of a calving ice-sheet model. In A. Berger, J. Imbrie, J. Hays, G. Kukla and B. Saltzman (eds), *Milankovitch and Climate*, D.

Reidel, Dordrecht, pp. 541–564.

Pollard, D., Ingersoll, A.P. and Lockwood, J.G. (1980) Response of a zonal climate ice-sheet model to the orbital perturbations during the Quaternary ice ages. *Tellus*, 32, 301–319.

Ponel, P. and Coope, G.R. (1990) Lateglacial and early Flandrian Coleoptera from La Taphanel, Massif Central, France: climatic and ecological implications. *Journal of Quaternary Science, 5*(3), 235–249.

Pons, A. and Reille, M.T. (1988) The Holocene and upper Pleistocene pollen record from Padul (Granada, Spain): a new study. *Palaeogeography, Palaeoclimatology, Palaeoecology, 66*, 243–263.

Porter, S.C. (1989) Some geological implications of average Quaternary glacial conditions. *Quaternary Research, 32*(3), 245–261.

Prell, W.L. and Kutzbach, J.E. (1987) Monsoon variability over the past 150,000 years. *Journal of Geophysical Research, 92*(D7), 8411–8425.

Ramanathan, V., Cicerone, R.J., Singh, H.B. and Kiehl, J.T. (1985) Trace gases and their potential role in climate change. *Journal of Geophysical Research, 90*(D3), 5547–5566.

Ramanathan, V., Cess, R.D., Harrison, E.F., Minnis, P., Barkstrom, B.R., Ahmad, E. and Hartmann, D. (1989) Cloud-radiative forcing and climate: results from the Earth Radiation Budget Experiment. *Science, 243*, 57–63.

Rampino, M.R. and Self, S. (1982) Historic eruptions of Tambora (1815), Krakatau (1883), and Agung (1963), their stratospheric aerosols and climatic impact. *Quaternary Research, 18*(2), 127–143.

Raper, S.C.B. and Wigley, T.M.L. (1991) Short-term global mean temperature and sea level change. In C.M. Goodess and J.P. Palutikof (eds), *Proceedings of the International Workshop on Future Climate Change and Radioactive Waste Disposal*, Nirex Safety Series, NSS/R257, pp. 203–213.

Raper, S.C.B., Wigley, T.M.L. and Warrick, R.A. (1992) Global sea level rise: Past and Future. In J.D. Milliman (ed), *Proceedings of the SCOPE Workshop on Rising Sea Level and Subsiding Coastal Areas, Bangkok, 1988*, John Wiley and Sons, Chichester, in press.

Raynaud, D., Chappellaz, J., Barnola, J.M., Korotkevich, Y.S. and Lorius, C. (1988) Climatic and CH_4 cycle implications of glacial-interglacial CH_4 change in the Vostok ice core. *Nature, 333*(6174), 655–657.

Rendell, H., Worsley, P., Green, F. and Parks, D. (1991) Thermoluminescence dating of the Chelford interstadial. *Earth and Planetary Science Letters, 103*(1–4), 182–189.

Revelle, R. (1983) Probable future changes in sea level resulting from increased atmospheric carbon dioxide. In National Academy of Sciences (ed.), *Changing Climate*, Report of the Carbon Dioxide Committee, Board of Atmospheric Sciences and Climate. National Academy Press, Washington, DC.

Revelle, R.R. and Waggoner, P.E. (1983) Effects of a carbon dioxide-induced climatic change on water supplies in the western United States. In National Academy of Sciences (ed.), *Changing Climate*, Report of the Carbon Dioxide Committee, Board of Atmospheric Sciences and Climate. National Academy Press, Washington, DC, pp. 419–431.

Ribes, E. (1990) Astronomical determinations of the solar variability. *Philosophical Transactions of the Royal Society of London A, 330*, 487–498.

Richmond, G.M. and Fullerton, D.S. (1986) Summation of Quaternary glaciations in the United States of America. *Quaternary Science Reviews, 5*, 183–196.

Rietmeijer, F.J.M. (1990) El Chichón dust a persistent problem. *Nature, 344*(6262), 114–115.

Rind, D. (1987) Components of the ice age circulation. *Journal of Geophysical*

Research, 92(D4), 4244–4281.

Rind, D. (1988) Dependence of warm and cold climate depiction on climate model resolution. *Journal of Climate, 1*(10), 965–997.

Rind, D., Peteet, D. and Kukla, G. (1989) Can Milankovitch orbital variations initiate the growth of ice sheets in a general circulation model? *Journal of Geophysical Research, 94*(D10), 12851–12871.

Rind, D., Chiou, E.W., Chu, W., Larsen, J., Oltmans, S., Lerner, J., McCormick, M.P. and McMaster, L. (1991) Positive water vapour feedback in climate models confirmed by satellite data. *Nature, 349*(6309), 500–503.

Robin, G. de Q. (1986) Changing the sea level. In B. Bolin, J. Jäger, B.R. Doos and R.A. Warrick (eds), *The Greenhouse Effect: Climatic Change and Ecosystems*, SCOPE Report 29, Wiley, New York, pp. 323–359.

Robinson, P.J. (1991) Scenarios of future climate for impact assessment. *The Environmental Professional, 13*, 8–15.

Robinson, P.J. and Finkelstein, P.L. (1991) The development of impact-oriented climate scenarios. *Bulletin of the American Meteorological Society, 72*(4), 481–490.

Rooth, C.G.H. (1990) Meltwater Younger Dryas upheld. *Nature, 343*(6260), 702.

Rose, J., Turner, C., Coope, G.R. and Bryan, M.D. (1980) Channel changes in a lowland river catchment over the last 13,000 years. In R.A. Cullingford, D.A. Davidson and J. Lewin (eds), *Timescales in Geomorphology*, Wiley, Chichester, pp. 159–175.

Rossignol-Strick, M. and Planchais, N. (1989) Climate patterns revealed by pollen and oxygen isotope records of a Tyrrhenian sea core. *Nature, 342*(6248), 413–416.

Röthlisberger, F. (1986) *10,000 Jahre Gletschergeschichte der Erde*. Verlag Sauerlander, Aarau.

Rousseau, D.D. and Puissegur, J.J. (1990) A 350,000-year climatic record from the loess sequence of Achenheim, Alsace, France. *Boreas, 19*(3), 203–216.

Rowlands, B.M. (1971) Radiocarbon evidence of the age of an Irish Sea glaciation in the Vale of Clwyd. *Nature, Physical Science, 230*, 9–10.

Royal Society (1990) The Earth's climate and variability of the Sun over recent millennia: geophysical, astronomical and archaeological aspects. *Philosophical Transactions of the Royal Society of London A, 330*, 399–685.

Ruddiman, W.F. and Kutzbach, J.E. (1989) Forcing of late Cenozoic Northern Hemisphere climate by plateau uplift in southern Asia and the American west. *Journal of Geophysical Research, 94(D15)*, 18409–18427.

Ruddiman, W.F. and Kutzbach, J.E. (1991) Plateau uplift and climatic change. *Scientific American, 264*(3), 42–50.

Ruddiman, W.F. and McIntyre, A. (1976) Northeast Atlantic paleoclimatic changes over the past 600,000 years. *Geological Society of America, Memoir 145*, 111–146.

Ruddiman, W.F. and McIntyre, A. (1981a) The North Atlantic Ocean during the last deglaciation. *Palaeogeography, Palaeoclimatology, Palaeoecology, 35*, 145–214.

Ruddiman, W.F. and McIntyre, A. (1981b) The mode and mechanism of the last deglaciation: oceanic evidence. *Quaternary Research, 16*(2), 125–134.

Ruddiman, W.F., Shackleton, N.J. and McIntyre, A. (1986) North Atlantic sea-surface temperatures for the last 1.1 million years. In C.P. Summerhayes and N.J. Shackleton (eds), *North Atlantic Palaeoceanography*, Geological Society Special Publication No. 21, pp. 155–173.

Ruddiman, W.F., Prell, W.L. and Raymo, M.E. (1989) Late Cenozoic uplift in southern Asia and the American west: rationale for general circulation modeling experiments. *Journal of Geophysical Research, 94*(D15), 18379–18391.

Saltzman, B. and Maasch, K.A. (1988) Orbital forcing and the Vostok ice core. *Nature, 333*(6169), 123–124.

Saltzman, B. and Sutera, A. (1984) A model of the internal feedback system involved in late Quaternary climatic variations. *Journal of Atmospheric Science,* *41*(5), 736–745.

Saltzman, B., Hansen, A.R. and Maasch, K.A. (1984) The Late Quaternary glaciations as the response of a three-component feedback system to earth-orbital forcing. *Journal of the Atmospheric Sciences, 41*(23), 3380–3389.

Santer, B. (1988) *Regional Validation of General Circulation Models,* Climatic Research Unit Research Publication, CRURP9.

Santer, B.D. and Wigley, T.M.L. (1990) Regional validation of means, variances, and spatial patterns in general circulation model control runs. *Journal of Geophysical Research, 95*(D1), 829–850.

Sarnthein, M., Winn, K., Duplessy, J.-C. and Fontugne, M.R. (1988) Global variations of surface ocean productivity in low and mid latitudes: influence on CO_2 reservoirs of the deep ocean and atmosphere during the last 21,000 years. *Palaeoceanography, 3,* 361–399.

Savoie, D.L. and Prospero, J.M. (1989) Comparison of oceanic and continental sources of non-sea-salt sulphate over the Pacific Ocean. *Nature, 339*(6227), 685–687.

Schlesinger, M.E. (1988) Negative or positive cloud optical depth feedback? *Nature, 335*(6188), 303–304.

Schlesinger, M.E. and Mitchell, J.F.B. (1987) Climate model simulations of the equilibrium climatic response to increased carbon dioxide. *Reviews of Geophysics, 25*(4), 760–798.

Schlesinger, M.E. and Oh, J.-H. (1988) A parameterization of the evaporation of rainfall. *Monthly Weather Review, 116*(10), 1887–1895.

Schlesinger, M.E. and Zhao, Z.-C. (1989) Seasonal climatic changes induced by doubled CO_2 as simulated by the OSU atmospheric GCM/mixed-layer ocean model. *Journal of Climate, 2*(5), 459–495.

Schöve, D.J. (1987) Sunspot cycles and weather history. In M.R. Rampino, J.E. Sanders, W.S. Newman and L.K. Koñigsson (eds), *Climate, History, Periodicity, and Predictability,* Van Nostrand Reinhold, New York, pp. 355–377.

Schwartz, S.E. (1988) Are global cloud albedo and climate controlled by marine phytoplankton? *Nature, 336*(6198), 441–445.

Sear, C.B., Kelly, P.M., Jones, P.D. and Goodess, C.M. (1987) Global surface temperature response to major volcanic eruptions. *Nature, 330*(6146), 365–367.

Seret, G., Dricot, E. and Wansard, G. (1990) Evidence for an early glacial maximum in the French Vosges during the last glacial cycle. *Nature, 346*(6283), 453–456.

Shackleton, N.J. (1967) Oxygen isotope analyses and Pleistocene temperatures reassessed. *Nature, 215*(5096), 15–17.

Shackleton, N.J. (1989) Palaeoclimate: deep trouble for climate change. *Nature, 342*(6250), 616–617.

Shackleton, N.J. and Imbrie, J. (1990) The $\delta^{18}O$ spectrum of oceanic deep water over a five-decade band. *Climatic Change, 16*(2), 217–230.

Shackleton, N.J. and Opdyke, N.D. (1973) Oxygen isotope and paleomagnetic stratigraphy of equatorial Pacific core V28-238: oxygen isotope temperatures and ice volumes on a 10^5 and 10^6 year scale. *Quaternary Research, 3*(1), 39–55.

Shackleton, N.J. and Opdyke, N.D. (1976) Oxygen-isotope and paleomagnetic stratigraphy of Pacific core V28-239 late Pliocene to latest Pleistocene. *Geological Society of America, Memoir 145,* 449–464.

Shackleton, N.J., Imbrie, J. and Hall, M.A. (1983) Oxygen and carbon isotope record of East Pacific core V19-30: implications for the formation of deep water in the late Pleistocene North Atlantic. *Earth and Planetary Science Letters, 65,* 233–244.

References

Shackleton, N.J., Imbrie, J. and Pisias, N.G. (1988) The evolution of oceanic oxygen-isotope variability in the North Atlantic over the past three million years. *Philosophical Transactions of the Royal Society of London B, 318*, 679–688.

Shaw, J. (1989) Drumlins, subglacial meltwater floods, and ocean response. *Geology, 17*, 853–856.

Shepard, F.P. (1963) Thirty-five thousand years of sea level. In T. Clements (ed.), *Essays in Marine Geology in Honour of K.O. Emery*, University of Southern California Press, Los Angeles.

Shine, K.P., Derwent, R.G., Wuebbles, D.J. and Morcrete, J- J. (1990) Radiative forcing of climate. In J.T. Houghton, G.J. Jenkins and J.J. Ephraums (eds), *Climate Change, the IPCC Scientific Assessment*, WMO/UNEP, Cambridge University Press, Cambridge, pp. 41–68.

Shinn, R.A. and Barron, E.J. (1989) Climate sensitivity to continental ice sheet size and configuration. *Journal of Climate, 2*(12), 1517–1537.

Short, D.A., Mengel, J.G., Crowley, T.J., Hyde, W.T. and North, G.R. (1991) Filtering of Milankovitch cycles by Earth's geography. *Quaternary Research, 35*(2), 157–173.

Shukla, J., Nobre, C. and Sellers, P. (1990) Amazon deforestation and climate change. *Science, 247*, 1322–1325.

Sibrava, V. (1986) Correlation of European glaciations and their relation to the deep-sea record. *Quaternary Science Reviews, 5*, 433–441.

Siegenthaler, U. and Wenk, T. (1984) Rapid atmospheric CO_2 variations and ocean circulation. *Nature, 308*(5960), 624–626.

Sissons, J.B. (1974) A late-glacial ice-cap in the central Grampians, Scotland. *Transactions of the Institute of British Geographers, 62*, 95–114.

Slade, D.H. (1990) A survey of informed opinion regarding the nature and reality of a 'global greenhouse warming' — an editorial. *Climatic Change, 16*(1), 1–4.

Slingo, A. (1990) Sensitivity of the Earth's radiation budget to changes in low clouds. *Nature, 343*(6253), 49–51.

Slingo, A. and Slingo, J.M. (1988) The response of a general circulation model to cloud longwave radiative forcing. I: Introduction and initial experiments. *Quarterly Journal of the Royal Meteorological Society, 114*(482 Part A), 1027–1062.

Smith, G.I. (1984) Paleohydrologic regimes in the Southwestern Great Basin, 0–3.2 My ago, compared with other long records of 'global' climate. *Quaternary Research, 22*(1), 1–17.

Snieder, R.K. (1985) The origin of the 100,000 year cycle in a simple ice age model. *Journal of Geophysical Research, 90*(D3), 5661–5664.

Sofia, S., O'Keefe, J., Lesh, J.R. and Endel, A.S. (1979) Solar constant: constraints on possible variations derived from solar diameter measurements. *Science, 204*, 1306–1308.

Sonett, C.P. and Finney, S.A. (1990) The spectrum of radiocarbon. *Philosophical Transactions of the Royal Society of London A, 330*, 413–426.

Souchez, R., Lemmens, M., Lorrain, R., Tison, J.L., Jouzel, J. and Sugden, D. (1990) Influence of hydroxyl-bearing minerals on the isotopic composition of ice from the basal zone of an ice sheet. *Nature, 345*(6272), 244–246.

Spelman, M.J. and Manabe, S. (1984) Influence of oceanic heat transport upon the sensitivity of a model climate. *Journal of Geophysical Research, 89*(C1), 571–586.

Stauffer, B., Hofer, H., Oeschger, H., Schwander, J. and Siegenthaler, U. (1984) Atmospheric CO_2 concentration during the last glaciation. *Annals of Glaciology, 5*, 160–164.

Stauffer, B., Lochbronner, E., Oeschger, H. and Schwander, J. (1988) Methane concentration in the glacial atmosphere was only half that of the preindustrial Holocene. *Nature*, *332*(6167), 812–814.

Stephenson, F.R. and Wolfendale, A.W. (eds) (1988) *Secular Solar and Geomagnetic Variations in the Last 10,000 Years*, NATO Advanced Science Institutes Series C, Mathematical and Physical Science 236, Kluwer Academic Publishers, Dordrecht.

Stevenson, A.C. and Moore, P.D. (1982) Pollen analysis of an interglacial deposit at West Angle, Dyfed, Wales. *New Phytologist*, *90*, 327–337.

Stocker, T.F. and Wright, D.G. (1991) Rapid transitions of the ocean's deep circulation induced by changes in surface water fluxes. *Nature*, *351*(6329), 729–732.

Stoker, M.S., Harland, R., Morton, A.C. and Graham, D.K. (1989) Late Quaternary stratigraphy of the northern Rockall Trough and Faeroe-Shetland Channel, northeast Atlantic Ocean. *Journal of Quaternary Science*, *4*(3), 211–222.

Stommel, H.M. (1957) The abyssal circulation. *Deep Sea Research*, *5*, 80–82.

Stothers, R.B. (1987) Do slow orbital periodicities appear in the record of early magnetic reversals. *Geophysical Research Letters*, *14*(11), 1087–1090.

Stouffer, R.J., Manabe, S. and Bryan, K. (1989) Interhemispheric asymmetry in climate response to a gradual increase of atmospheric CO_2. *Nature*, *342*(6250), 660–662.

Street-Perrott, F.A. (1991) GCM modelling of palaeoclimates. In C.M. Goodess and J.P. Palutikof (eds), *Proceedings of the International Workshop on Future Climate Change and Radioactive Waste Disposal*, Nirex Safety Series, NSS/R257, pp. 167–178.

Street-Perrott, F.A. and Perrott, R.A. (1990) Abrupt climatic fluctuations in the tropics: the influence of Atlantic Ocean circulation. *Nature*, *343*(6259), 607–612.

Stringer, C.B., Currant, A.P., Schwarcz, H.P. and Collcutt, S.N. (1986) Age of Pleistocene faunas from Bacon Hole, Wales. *Nature*, *320*(6057), 59–62.

Stuiver, M. and Becker, B. (1986) High-precision decadal calibration of the radiocarbon time scale, AD 1950–2500 BC. *Radiocarbon*, *28*, 863–910.

Stuiver, M. and Quay, P.D. (1980) Changes in atmospheric carbon-14 attributed to a variable sun. *Science*, *207*, 11–19.

Stuiver, M., Pearson, G.W. and Braziunas, T. (1986) Radiocarbon age calibration of marine samples back to 9000 cal yr BP. *Radiocarbon*, *28*, 980–1021.

Stuiver, M., Braziunas, T.F., Becker, B. and Kromer, B. (1991) Climatic, solar, oceanic, and geomagnetic influences on late-glacial and Holocene atmosphere C^{14}/C^{12} change. *Quaternary Research*, *35*(1), 1–24.

Suess, H.E. (1980) The radiocarbon record in tree rings of the last 8000 years. *Radiocarbon*, *22*, 200–209.

Sutcliffe, A.J., Lord, T.C., Harmon, R.S., Ivanovich, M., Rae, A. and Hess, J.W. (1985) Wolverine in northern England at about 83,000 yr BP: faunal evidence for climatic change during isotope stage 5. *Quaternary Research*, *24*(1), 73–86.

Sutherland, D.G., Ballantyne, C.K. and Walker, M.J.C. (1984) Late Quaternary glaciation and environmental change on St Kilda, Scotland, and their palaeoclimatic significance. *Boreas*, *13*(3), 261–272.

Teller, J.T. (1987) Proglacial lakes and the southern margin of the Laurentide ice sheet. In W.F. Ruddiman and H.E. Wright (eds), *North America and Adjacent Oceans during the Last Deglaciation*, Geological Society of America, pp. 39–69.

Teller, J.T. (1990) Volume and routing of late-glacial runoff from the southern Laurentide ice sheet. *Quaternary Research*, *34*(1), 12–23.

Thomas, R.H. (1986) Future sea level rise and its early detection by satellite remote sensing. In J.G. Titus (ed.) *Effects of Changes in Stratospheric Ozone and Global*

References

Climate, Volume 4: Sea Level Rise, US Environmental Protection Agency.

Thorarinsson, S. (1940) Present glacier shrinkage and eustatic changes in sea level. *Geografiska Annaler*, *22*, 131–159.

Tinsley, B.A. (1988) The solar cycle and the QBO influences on the latitude of storm tracks in the North Atlantic. *Geophysical Research Letters*, *15*(5), 409–412.

Tipping, R.M. (1989) Long-distance transported Pinus pollen as a possible chronostratigraphic marker in the Scottish early postglacial. *Boreas*, *18*(4), 333–341.

Tooley, M.J. (1974) Sea-level changes during the last 9000 years in north-west England. *Geographical Journal*, *140*, 18–42.

Tooley, M.J. (1989) Global sea levels: floodwaters mark sudden rise. *Nature*, *342*(6245), 20–21.

Trewartha, G.T. (1968) *An Introduction to Climate* (4th edn). McGraw-Hill, London.

Tricot, C., Gallée, H., Fichefet, T., Marsiat, I. and Berger, A. (1989) A simulation of the long-term variations of the global ice volume over the past 122,000 years: a test of the astronomical theory. In J. Lenoble and J.F. Geleyn (eds), *IRS '88: Current Problems in Atmospheric Radiation*, A. Deepak Publishing, pp. 338–341.

Trupin, A. and Wahr, J. (1990) Spectroscopic analysis of global tide gauge sea level data. *Geophysical Journal International*, *100*(3), 441–453.

Tsonis, A.A. and Elsner, J.B. (1988) The weather attractor over very short timescales. *Nature*, *333*(6173), 545–547.

Tsonis, A.A. and Elsner, J.B. (1989) Chaos, strange attractors, and weather. *Bulletin of the American Meteorological Society*, *70*(1), 14–23.

Tsonis, A.A. and Elsner, J.B. (1990) Multiple attractors, fractal basins and longterm climate dynamics. *Beiträge zur Physik der Atmosphäre*, *63*(3/4), 171–176.

Turner, C. and Hannon, G.E. (1988) Vegetational evidence for late Quaternary climatic changes in southwest Europe in relation to the influence of the North Atlantic Ocean. *Philosophical Transactions of the Royal Society of London B*, *318*, 451–485.

Turner, C. and West, R.G. (1968) The subdivision and zonation of interglacial periods. *Eiszeitalter und Gegenwart*, *19*, 93.

Turon, J.-L. (1984) Direct land/sea correlations in the last interglacial complex. *Nature*, *309*(5970), 673–676.

Tushingham, M. and Peltier, W.R. (1991) Ice-3G: A new global model of Late Pleistocene deglaciation based upon geophysical predictions of post-glacial relative sea level change. *Journal of Geophysical Research*, *96*, 4497–4523.

Vandenberghe, J. and Kasse, C. (1989) Periglacial environments during the early Pleistocene in the southern Netherlands and northern Belgium. *Palaeogeography, Palaeoclimatology, Palaeoecology*, *72*(1–2), 133–139.

Van der Hammen, T., Wijmstra, T.A. and Zagwijn, W.H. (1971) The floral record of the late Cenozoic of Europe. In K.K. Turekian (ed.), *The Late Cenozoic Glacial Ages*, Yale University Press, pp. 391–424.

Van der Veen, C.J. (1987) Ice sheets and the CO_2 problem. *Surveys in Geophysics*, *9*(1), 1–42.

Van der Veen, C.J. (1988) Projecting future sea level. *Surveys in Geophysics*, *9*(3–4), 389–418.

van Loon, H. and Labitzke, K. (1988) Association between the 11-year solar cycle, the QBO, and the atmosphere. Part II: surface and 700 mb in the Northern Hemisphere in winter. *Journal of Climate*, *1*(9), 905–920.

Vasari, Y. (1977) Radiocarbon dating of the lateglacial and early Flandrian vegetational succession in the Scottish Highlands and the Isle of Skye. In J.M.

Gray and J.J. Lowe (eds), *Studies in the Scottish Lateglacial Environment*, Pergamon Press, Oxford, pp. 143–162.

Venne, D.E. and Dartt, D.G. (1990) An examination of possible solar cycle–QBO effects in the Northern Hemisphere troposphere. *Journal of Climate*, 3(2), 272–281.

Vernekar, A.D. (1968) Long-period global variations of incoming solar radiation. In *Research on the Theory of Climate II*. Report on the Travelers Research Center Inc., Hartford, Connecticut.

Vernekar, A.D. (1972) *Long-period Global Variations of Incoming Solar Radiation*. Meteorological Monographs, 12. American Meteorological Society, Boston.

Vogel, J.S., Cornell, W., Nelson, D.E. and Southon, J.R. (1990) Vesuvius/ Avellino, one possible source of seventeenth century BC climatic disturbances. *Nature*, 344(6266), 534–537.

Walker, D. (1966) The Late Quaternary history of the Cumberland lowland. *Philosophical Transactions of the Royal Society of London B*, 251, 1–210.

Walker, M.J.C. (1980) Late-glacial history of the Brecon Beacons, south Wales. *Nature*, 287(5778), 133–135.

Walker, M.J.C. (1982) The late-glacial and early Flandrian deposits at Traeth Mawr, Brecon Beacons, south Wales. *New Phytologist*, 90, 177–194.

Walker, M.J.C. (1984) Pollen analysis and Quaternary research in Scotland. *Quaternary Science Reviews*, 3, 369–404.

Walker, M.J.C. and Harkness, D.D. (1990) Radiocarbon dating the Devensian late-glacial in Britain — new evidence from Llanilid, south Wales. *Journal of Quaternary Science*, 5(2), 135–144.

Wang, Y., Evans, M.E., Rutter, N. and Ding, Z. (1990) Magnetic susceptibility of Chinese loess and its bearing on paleoclimate. *Geophysical Research Letters*, 17(13), 2449–2451.

Warrick, R.A. (1986) Climatic change and sea level rise. *Climate Monitor*, 15(2), 39–44.

Warrick, R.A. and Oerlemans, H. (1990) Sea level rise. In J.T. Houghton, G.J. Jenkins and J.J. Ephraums (eds), *Climate Change, the IPCC Scientific Assessment*, WMO/UNEP, Cambridge University Press, Cambridge, pp. 257–282.

Warrilow, D.A. and Buckley, E. (1989) The impact of land surface processes on the moisture budget of a climate model. *Annales Geophysicae*, 7(5), 439–450.

Washington, W.M. and Meehl, G.A. (1984) Seasonal cycle experiment on the climate sensitivity due to a doubling of CO_2 with an atmospheric general circulation model coupled to a simple mixed-layer ocean model. *Journal of Geophysical Research*, 89(D6), 9475–9503.

Washington, W.M. and Meehl, G.A. (1989) Climate sensitivity due to increased CO_2: experiments with a coupled atmosphere and ocean general circulation model. *Climate Dynamics*, 4, 1–38.

Washington, W.M. and Parkinson, C.L. (1986) *An Introduction to Three-dimensional Climate Modelling*. University Science Books, Oxford.

Watson, E. (1977) The periglacial environment: the periglacial environment of Great Britain during the Devensian. *Philosophical Transactions of the Royal Society of London B*, 280, 183–198.

Watson, R.T., Rodhe, H., Oeschger. H. and Siegenthaler, U. (1990) Greenhouse gases and aerosols. In J.T. Houghton, G.J. Jenkins and J.J. Ephraums (eds), *Climate Change, the IPCC Scientific Assessment*, WMO/UNEP, Cambridge University Press, Cambridge, pp. 1–40.

Watts, R.G. and Hayder, E. (1984) A two dimensional, seasonal, energy balance climate model with continents and ice sheets: testing the Milankovitch theory.

Tellus, *36*(2), 120– 131.

Webb, T., III and Wigley, T.M.L. (1985) What past climates can indicate about a warmer world. In M.C. MacCracken and F.M. Luther (eds), *Projecting the Climatic Effects of Increasing Carbon Dioxide*, DOE/ER-0237, US Department of Energy, Carbon Dioxide Research Division, Washington, DC, pp. 237–257.

Weertman, J. (1976) Milankovitch solar radiation variations and ice age ice sheet sizes. *Nature*, *261*(5555), 17–20

Weidick, A. (1984) Review of glacier changes in West Greenland. *Zeitschrift fur Gletscherfunde und Glazialgeologie*, *21*, 301–309.

Weiss, N.O. (1990) Periodicity and aperiodicity in solar magnetic activity. *Philosophical Transactions of the Royal Society of London A*, *330*, 617–626.

West, R.G. (1988) The record of the cold stages. *Philosophical Transactions of the Royal Society of London B*, *318*, 505–522.

Wetherald, R.T. and Manabe, S. (1986) An investigation of cloud cover change in response to thermal forcing. *Climatic Change*, *8*(1), 5–23.

Weyl, P.K. (1968) The role of the oceans in climatic change. A theory of the ice ages. *Meteorological Monographs*, *8*, 37–62.

Wigley, T.M.L. (1981) Climate and paleoclimate: what we can learn about solar luminosity variations. *Solar Physics*, *74*, 435–471.

Wigley, T.M.L. (1983) The preindustrial carbon dioxide level. *Climatic Change*, *5*(4), 315–320.

Wigley, T.M.L. (1988a) The climate of the past 10,000 years and the role of the sun. In R.F. Stephenson and A.W. Wolfendale (eds), *Secular Solar and Geomagnetic Variations in the Last 10,000 years*, Kluwer Academic Publishers, Dordrecht, pp. 209–224.

Wigley, T.M.L. (1988b) Future CFC concentrations under the Montreal Protocol and their greenhouse-effect implications. *Nature*, *335*(6188), 333–335.

Wigley, T.M.L. (1989) Possible climate change due to SO_2-derived cloud condensation nuclei. *Nature*, *339*(6223), 365– 367.

Wigley, T.M.L. and Barnett, T.P. (1990) Detection of the greenhouse effect in the observations. In J.T. Houghton, G.J. Jenkins and J.J. Ephraums (eds), *Climate Change, the IPCC Scientific Assessment*, WMO/UNEP, Cambridge University Press, Cambridge, pp. 239–256.

Wigley, T.M.L. and Kelly, P.M. (1990) Holocene climatic change: ^{14}C wiggles and variations in solar irradiance. *Philosophical Transactions of the Royal Society of London A*, *330*, 547–560.

Wigley, T.M.L. and Raper, S.C.B. (1987) Thermal expansion of sea water associated with global warming. *Nature*, *330*(6144), 127–131.

Wigley, T.M.L. and Raper, S.C.B. (1990) Natural variability of the climate system and detection of the greenhouse effect. *Nature*, *344*(6264), 324–326.

Wigley, T.M.L. and Raper, S.C.B. (1992) Future trends in global-mean temperature and thermal-expansion-related sea level rise. In R.A. Warrick, E.M. Barrow and T.M.L. Wigley (eds), *Climate and Sea Level Change: Observations, Projections and Implications*, Cambridge University Press, Cambridge, in press.

Wigley, T.M.L. and Santer, B.D. (1990) Statistical comparison of spatial fields in model validation, perturbation, and predictability experiments. *Journal of Geophysical Research*, *95*(D1), 851–865.

Wigley, T.M.L., Jones, P.D. and Kelly, P.M. (1986) Empirical climate studies: warm world scenarios and the detection of climatic change induced by radiatively active gases. In B. Bolin, B.R. Doos, J. Jäger and R.A. Warrick (eds), *The Greenhouse Effect, Climatic Change, and Ecosystems*, SCOPE 29, Wiley, New York, pp. 271–322.

Wigley, T.M.L., Jones, P.D., Briffa, K.R. and Smith, G. (1990) Obtaining sub-grid-scale information from coarse-resolution general circulation model output. *Journal of Geophysical Research*, *95*(D2), 1943–1953.

Wigley, T.M.L., Santer, B.D., Schlesinger, M.E. and Mitchell, J.F.B. (1992) Developing climate scenarios from equilibrium GCM results. *Climatic Change*, submitted.

Wijmstra, T.A. (1969) Palynology of the first 30 meters of a 120 m deep section in northern Greece. *Acta Botanica Neerlandica*, *18*, 511–527.

Wilks, D.S. (1989) Statistical specification of local surface weather elements from large-scale information. *Theoretical and Applied Climatology*, *40*(3), 119–134.

Williams, L.D. and Wigley, T.M.L. (1983) A comparison of evidence for late Holocene summer temperature variations in the Northern Hemisphere. *Quaternary Research*, *20*(3), 286–307.

Williams, R.B.G. (1975) The British climate during the last glaciation; an interpretation based on periglacial phenomena. In A. F. Wright and F. Moseley (eds), *Ice Ages, Ancient and Modern*, Seel House Press, Liverpool, pp. 95–120.

Wilson, C.A. and Mitchell, J.F.B. (1987) A doubled CO_2 climate sensitivity experiment with a global climate model including a simple ocean. *Journal of Geophysical Research*, *92*(D11), 13315–13343.

Winograd, I.J., Szabo, B.J., Coplen, T.B. and Riggs, A.C. (1988) A 250,000 year climatic record from Great Basin vein calcite: implications for Milankovitch theory. *Science*, *242*, 1275–1280.

Woillard, G. (1979) Abrupt end of the last interglacial s.s. in north-east France. *Nature*, *281*(5732), 558–562.

Woillard, G.M. and Mook, W.G. (1982) Carbon-14 dates at Grande Pile: correlation of land and sea chronologies. *Science*, *215*, 159–161.

Zardini, D., Raynaud, D., Scharffe, D. and Seiler, W. (1989) N_2O measurements of air extracted from Antarctic cores: implication on atmospheric N_2O back to the last glacial-interglacial transition. *Journal of Atmospheric Chemistry*, *8*(2), 189–201.

Index